KT-485-230

How to use this book

The Animals. All Britain's native wild animals – including mammals, reptiles and amphibians – and all the commoner continental species, are included in this book. Also covered are animals which have been introduced to Europe and have become firmly established. Only the entirely marine animals are excluded.

The Illustrations. Every species is illustrated in colour and is shown, in most cases, in the habitat it normally occupies. Where appropriate, males, females, juveniles and seasonal or other variations are all shown.

On pages 110-121 the various tracks and signs that mammals leave behind them are illustrated. These include footprints, droppings, feeding signs (such as tooth and claw marks) and teeth, bones and antlers.

The Text. Each species is described on the page opening on which it is illustrated. Measurements refer to the length of a normal adult (the larger sex if there is a difference) from the end of its snout to the tip of its tail.

The introduction discusses the different types of animals included in the book, their similarities and their differences, and provides some valuable information about finding and watching wild animals. The illustrated contents key shows typical members of all the different groups covered by the book, enabling you to turn rapidly to the section you want.

At the end of the book, on pages 122-125, is an account of the distribution of animals (especially in Britain) to provide a guide to what animals you should look for in particular areas.

Finding Wild Animals. Most wild animals are secretive and many of them are nocturnal. Apart from a few common species, they can be difficult to find. Watching them thus requires great patience, and the ability to keep still and quiet. Remember, animals can hear, see or smell you long before you can see them.

It is often easier to find evidence of an animal's presence than it is to see the animal itself. Use the tracks and signs illustrations, together with the section on the distribution of animals, to find out what sort of wild animals are present in any particular area. You will then have a better idea of what to look for. In time, your detective skills will become second nature and you will find evidence of a greater richness and diversity of animal life than you might have expected.

A HANDGUIDE TO THE

WILD ANIMALS

OF BRITAIN AND EUROPE

Denys Ovenden,
Gordon Corbet and Nicholas Arnold

Designed by **Hermann Heinzel**

TREASURE PRESS

ACKNOWLEDGEMENTS

This title first appeared as the *Collins Handguide to the Wild Animals of Britain and Europe* and Treasure Press gratefully acknowledge the co-operation of William Collins Sons & Co Ltd who gave permission for this edition to be published.

First published in Great Britain in 1979 by
William Collins Sons & Co Ltd

This edition published in 1985 by
Treasure Press
Michelin House
81 Fulham Road
London SW3 6RB

Reprinted 1988, 1989

ISBN 1 85051 050 4

Printed in Portugal by Oficinas Gráficas ASA

Contents

Where more than one animal is listed the illustration shows a species typical of the group

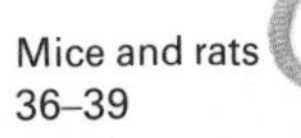

Bear 40
Wolves and foxes 41–43
Martens 50–51
Stoat, Weasel and Polecat 46–47
Otter and Mink 48–49
Wolverine 51
Raccoon and Raccoon dog 45
Badger 44–45
Wild cat 52–53
Lynx 52–53
Seals 54–55
Genet 53

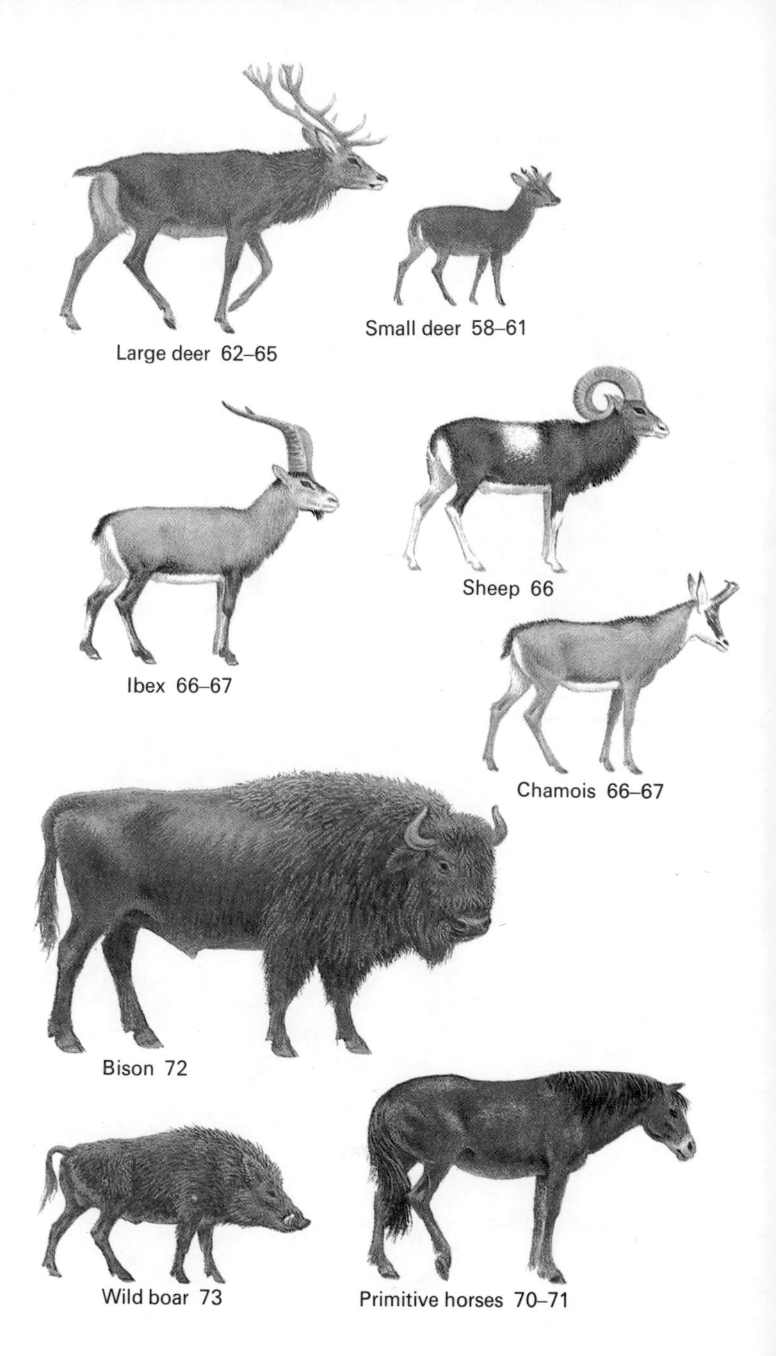
Large deer 62–65
Small deer 58–61
Ibex 66–67
Sheep 66
Chamois 66–67
Bison 72
Wild boar 73
Primitive horses 70–71

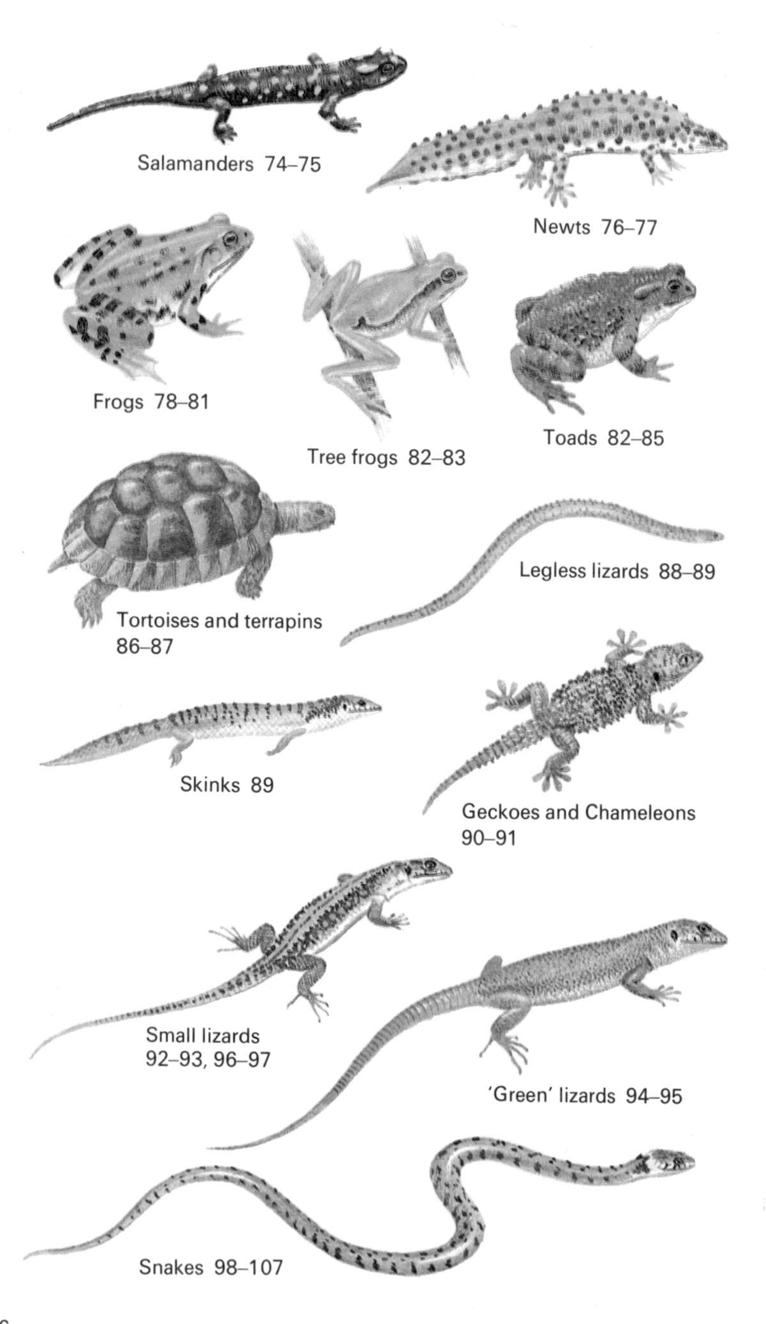
Salamanders 74–75
Newts 76–77
Frogs 78–81
Tree frogs 82–83
Toads 82–85
Tortoises and terrapins
86–87
Legless lizards 88–89
Skinks 89
Geckoes and Chameleons
90–91
Small lizards
92–93, 96–97
'Green' lizards 94–95
Snakes 98–107

Introduction

The continent of Europe offers a very wide range of climate, from near permanent frost in the north of Scandinavia to the baking heat on the frost-free islands of the Mediterranean. As the map below shows, the land surface is just as varied, with vast conifer forests of larch, spruce and pine,

Key to vegetation zones:

 Mediterranean scrub

 Coniferous forest

 Deciduous forest and grassland

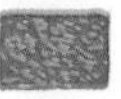 Steppe and desert margin

 Grassland

Alpine and tundra

hardwood forests of oak and beech, flat plains, sandy heaths and marshes. Much of this landscape is the result of man's intervention, for there was a time when Europe was largely forest, and open ground was found only above the tree line on the mountains and in the wet lands and marshes bordering the great rivers. Primitive man had a tremendous variety of wild animals to hunt. They included mammoth, woolly rhinoceros, cave bear and a host of others, but climatic change and the increase of human population caused many of the larger species to become extinct. The last of the aurochs, the great wild ox, vanished in Europe in the 17th century, and in Britain the last big carnivore, a wolf, was killed in Scotland in 1746.

This book is an introduction to the animals that remain. For the most part they are small, living on the ground or close to it, able to hide in rock crevices, beneath roots, in water or on the sheer crags of the Alps and the Pyrenees. Yet there is still a wide variety of animal life in Europe. Between the European bison and the minute pigmy shrew, smaller even than the Bison's footprint, there is a remarkable range of size, shape, colour and way of life, a range that is made even wider if we include the reptiles and amphibians.

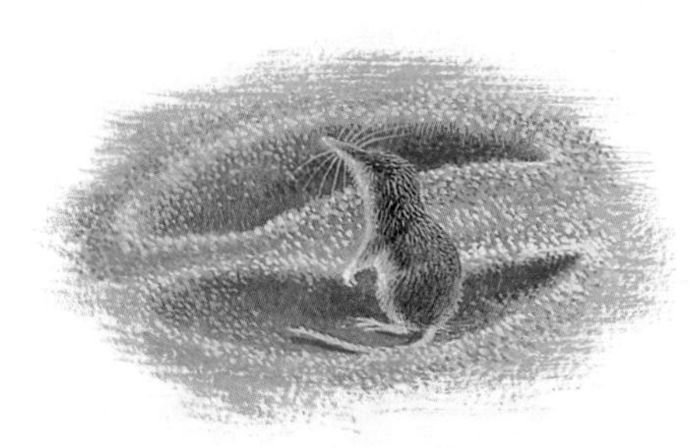

For most people, the words 'wild animals' bring to mind a deer, a fox, an otter or a badger. So it may seem strange that this book includes such things as frogs, snakes and lizards. Yet these are wild, and they are certainly animals. What is the difference between them?

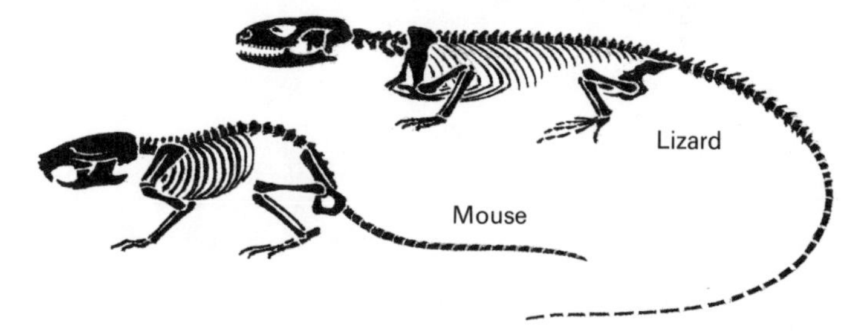

If you take a lizard and a mouse down to the basic framework of bones – the skeleton – you will find a general resemblance. They both have a backbone made up of a number of small bones or vertebrae. Hence we call them *vertebrates*, or animals with backbones. However, the mouse, in common with all fur- or hair-covered animals in Europe, is a mammal. This is the name given to all animals that feed their infant young with milk from teats or mammary glands, usually under the body. In addition, mammals are warm-blooded, that is to say, their bodies are usually kept at a steady warm temperature, whatever the external conditions. However, this pattern is complicated by the process of hibernation, whereby some small animals, dormice for instance, pass the winter in a deep sleep, with a fall in body temperature that would be enough to kill a human being. Mammals give birth to well-developed, more or less active young, depending on the kind, or species, of animal. Young mice are born blind, hairless and largely helpless. Young deer are born fully haired and sighted, and within minutes can run fast enough to reach concealment.

Reptiles and amphibians present a very different picture. Reptiles are usually dry-skinned, and clad in a layer of scales. In most species the scales overlap. In contrast, all European amphibians lack scales, and their skin is usually moist. Most reptiles lay eggs. These are either hard or soft shelled, and are often buried in the ground or in piles of rotting vegetation. A few species, such as the Viviparous lizard, may give birth to fully developed young which reach the hatching stage while the eggs are

Frog: an amphibian

Lizard: a reptile

still inside the mother's body. The babies are active little replicas of their parents.

Amphibians typically lay their jelly-covered eggs in water. These eggs hatch into a completely aquatic larval, or tadpole, form. This eventually changes into a miniature of the adult. (See pp 108–9.) As with some reptiles, a few species of European salamanders may produce fully developed young.

Finally, reptiles and amphibians are generally called cold-blooded animals. More accurately they are temperature variable animals, for within certain limits their temperature alters in relation to their surroundings. To avoid extremes of body temperature, they will seek shade and protection from excessive heat and in extreme cold they will seek winter quarters that are as free from frost as possible, either in burrows, tree stumps, deep mud or crevices in walls and rocks.

The scope of this book

All major groups of mammals, reptiles and amphibians are included except for the entirely marine mammals that never come ashore voluntarily – the whales, dolphins and porpoises – and the marine turtles which only do so in warm regions to lay their eggs. Nearly all the more distinctive kinds are illustrated and described. Species not illustrated are generally similar to illustrated ones and often cannot be distinguished without detailed study.

Finding animals

In any one area the number of different kinds of land or 'terrestrial' vertebrates may be as many as 60, for example in some of the richer parts of the Mediterranean coast; or as few as five on the tundra of the far north. But the casual observer will only see a small proportion of them. He may notice the odd rabbit, squirrel, lizard or frog, but to get a real idea of the diversity of animals in an area needs the skills of a good detective – patience, curiosity and sensitivity to the slightest clues of sound, sight or smell which can reveal a whole world of animals with which we often unknowingly share our country. You will see far more by choosing a spot to watch from, and keeping still and quiet. If you must move, do so slowly and quietly. Remember, animals can hear, see or smell you long before you have found them.

Roe deer

Mammals

Many mammals are nocturnal and therefore difficult to see. However, others are diurnal and by far the best time to see them is in the early morning or at dusk. For example, deer will often feed in the open at dawn, retiring to the cover of woodland when it becomes warm or when human activity increases. Looking for droppings will often show you the best places to watch, and there are many other signs of animal presence – chewed bark, eaten tips of shoots on young trees and tracks left in snow or mud. Tracks and signs are dealt with in more detail on pages 110–19. Binoculars are very helpful. They allow you to watch from a distance, without disturbing the animals. Remember that deer, like most mammals, have very sensitive noses – choose a spot down-wind from the place where you expect to see them. Rabbits can be watched in exactly the same way. During such a dawn watch you may also be lucky enough to see one of the more strictly nocturnal animals getting home late, perhaps a fox or a badger. This goes equally for the water animals, such as water vole, musk rat, coypu, or even the more elusive carnivores, like otter or mink.

Squirrels are about the easiest of all wild animals to see: they feel safe enough in the tree-tops not to be greatly bothered by human beings walking around below. Marmots and ground squirrels, except where they have become tame through artificial feeding, are more wary and need the 'wait and watch' approach. Seals, in spite of their size, are easy to overlook

when they are lying out on a rocky shore. A careful search with binoculars will often reveal a group of them where the naked eye sees only rocks.

The small rodents and shrews are particularly difficult to see. Many come out only at night but even the diurnal ones generally stick to dense cover. One kind, the bank vole, can often be watched by day amongst bushes or fallen trees in woodland. Other rodents can sometimes be seen at night by regularly putting down bait, such as seeds of any kind, at a suitable spot. Use a torch, if possible with a red glass. Shrews and voles can sometimes be found under logs (which should always be carefully replaced). On rocky coasts beaches littered with driftwood and seaweed are good habitats for shrews. The seaweed flies and sandhoppers provide them with an abundant food supply.

Bats in flight are easy to see but very difficult to identify. For their daytime roost or their winter hibernation bats occupy a variety of sites, from hollow trees to roof-spaces, cellars and caves. They are very vulnerable to disturbance: each time a bat is awakened it loses some of the fat reserves that are essential to sustain it through the winter. It therefore is most important, especially in winter, to disturb them as little as possible. Even the lower squeaks of bats are pitched at a very high frequency. If you are a man, and over 40, the chances are that you will have lost the ability to pick up sounds in this range.

Reptiles and amphibians

Few traces are left by reptiles, though the few that can be found are useful indicators, such as cast or 'sloughed' snake skins. Lizards often lie out on the same stone each day when basking in the sun. Such a stone is likely to be covered with their droppings. These are easily mistaken for bird droppings, being dark at one end and whitish at the other. Many reptiles are conservative in their habits, and once you have found them there is every chance that they will be found in the same place, or within a metre or so, on successive days. However, there are exceptions to this. The adder, for example, tends to shift its quarters after mating, frequently by a kilometre or more, but come spring and it will be found back at the previous year's courtship ground.

In general, reptiles and amphibians are much easier to approach than most mammals and it is often possible to get near enough to examine them in detail. Nevertheless, binoculars are still useful, particularly ones that can be focused at short distances. Most species usually sleep through the winter but the spring, when they come out of hiding and begin courtship, is a good time to look for them. In the summer they become more retiring and more difficult to find. As with mammals, early morning searches are most productive for seeing species that are regularly active by day but salamanders and some frogs and toads are found most easily after dusk by searching with a broad-beamed lamp; rainy evenings are best for this. Although many reptiles and amphibians may be encountered in a variety of habitats, in the spring the latter concentrate in large numbers at their breeding ponds and are very easy to see there. At this time especially, frogs and toads can be located by their voices. Each species has its own distinctive call, ranging from the echoing croak of the Marsh frog to the soft, mournful piping of the Fire-bellied toad and, with practice, it may be possible to distinguish up to half a dozen species in a single pond by this means. Most other reptiles and amphibians have weak voices or, more usually, none at all, but they may attract attention to themselves by the noises made as they move about. The continuous rustling of a tortoise ploughing through dense herbage can soon be recognized as different from the intermittent scrabbling of a foraging lizard.

Because they can be approached closely, it is tempting to try to catch reptiles and amphibians but they are delicate animals and even slight damage may seriously reduce their chances of survival. A lizard will shed its tail if grasped by it and, although the animal can grow a new one, it will be at a serious disadvantage while doing so, especially since the process requires a great deal of protein. If handling cannot be avoided it should be done with great care and amphibians should be held only with wet hands to protect their soft, usually moist skins. It goes without saying that venomous snakes should not be handled in any circumstances.

△ **HEDGEHOG** *(Erinaceus europaeus).* Curled up into a tight ball and protected by its spiny coat, the hedgehog is safe from all but the most determined badger or fox. Found throughout Europe, except for the northern half of Scandinavia, hedgehogs feed at night, searching especially for earthworms, slugs, snails and insects, and can often be found on the damp grass of a lawn, sports ground or water meadow, wherever suitable cover for nesting is close at hand. They spend the day in a nest of leaves, in woodland or in a hedgerow or garden. In winter they hibernate. Body temperature drops to between 2° and 6°C and breathing slows to as little as ten per minute. Hedgehogs of eastern Europe differ from western animals in having white breasts instead of brown. Length 25cm.

ALGERIAN HEDGEHOG
◁ *(Erinaceus algirus).* The hedgehogs on the Balearic Islands and Malta belong to this North African species. A few live on the Mediterranean coasts of France and Spain. They look and behave much like the European hedgehogs but have a much wider spine-free 'parting' on the crown of the head, wide enough to insert a pencil, and they do not hibernate. Length 23cm.

PORCUPINE *(Hystrix cristata).* Porcupines look rather like large hedgehogs but, in fact, the two species have little in common apart from their spines. Porcupines are rodents and live in deep burrows, emerging at night to feed on bulbs, roots and bark. They can be very destructive to crops in Sicily and the other parts of Italy where they are found. The quills provide very effective protection. If a predator is rash enough to worry a porcupine, the latter will turn its back and run – not away, but backwards, driving its quills into its adversary's neck. The quills are not barbed and they cannot be shot out as is often supposed, but they are nevertheless formidable weapons. At the tip of the tail is a bunch of very peculiar quills. They are like very elongate, hollow goblets on slender stems and when the tail is shaken the quills produce a loud rattle not unlike that produced by a rattlesnake's tail. Length 70cm.

△

MOLE *(Talpa europaea).* Mounds of soil on lawns or meadows are the most familiar signs of moles. The animals themselves remain permanently underground, each in its own system of tunnels, which it digs with enormous spade-like front feet. They feed mainly on earthworms and in winter stores of worms may be accumulated in underground chambers, each one immobilized by a bite in the head. The mole is not totally blind, as is often thought, but its sight is extremely poor and the eyes are small. Length 15cm.

PYGMY SHREW *(Sorex minutus).* This tiny shrew is a voracious consumer of insects, which it hunts day and night amongst the litter of woodland or in long grass. The pygmy shrew must feed every few hours to keep alive. Yet it can survive even the coldest of winters without hibernating. Length 9cm.

COMMON SHREW *(Sorex araneus).* The most abundant and widespread shrew in northern Europe. They are more often heard than seen – an exceedingly high-pitched squeaking coming from grass or a hedgerow indicates that two shrews have met in a runway. They are aggressive little animals and most encounters tend to be hostile. Length 12cm.

GREATER WHITE-TOOTHED SHREW *(Crocidura russula).* Four species of white-toothed shrew are common in southern Europe. The 'caravan' behaviour seen here occurs when the nest is disturbed. Normally young shrews are fully grown by the time they leave the nest. Length 11cm.

WATER SHREW *(Neomys fodiens).* Water shrews are found mainly beside water. They can swim and dive expertly, hunting mainly for aquatic insects. Their saliva contains a poison capable of paralysing prey but too weak to be any danger to ourselves. Length 14cm.

PYGMY SHREW
COMMON SHREW
GREATER WHITE-TOOTHED SHREW
WATER SHREW

LESSER HORSESHOE BAT

DAUBENTON'S BAT

WHISKERED BAT

LARGE MOUSE-EARED BAT

LESSER HORSESHOE BAT *(Rhinolophus hipposideros).* The smallest and most widespread of the five kinds of horseshoe bat found in Europe with a complex series of fleshy lobes around the nostrils, the lower-most of which is shaped like a horseshoe. These 'nose-leaves' serve to direct the pulses of ultrasonic sound that the bats use as a form of radar to find their way around and to locate their insect prey. Like all European bats, they hibernate, sometimes in large colonies. Length 7cm.

DAUBENTON'S BAT *(Myotis daubentoni).* Daubenton's bats, also known as Water bats, are usually found near water. They emerge fairly early in the evening and may be seen hawking for insects over rivers and lakes. They spend the day in a hollow tree, a cave or even a small crevice amongst rocks or stonework. Like most European bats, they lack nose-leaves because they emit the pulses of sound used in echo-location through the mouth rather than the nostrils. Length 8cm.

WHISKERED BAT *(Myotis mystacinus).* One of the smallest of European bats, they hibernate mainly in caves, but tend to be solitary, clinging closely against the wall or in a small crevice. Length 8cm.

LARGE MOUSE-EARED BAT *(Myotis myotis).* With a wingspan of up to 45cm these are amongst the largest bats in Europe. Individuals marked with numbered rings on their wings have been recovered as far as 200km away from the place where they have been marked. Mouse-eared bats often live in large colonies, sometimes in houses, moving from the attic in summer to the cellar in winter. Length 12cm.

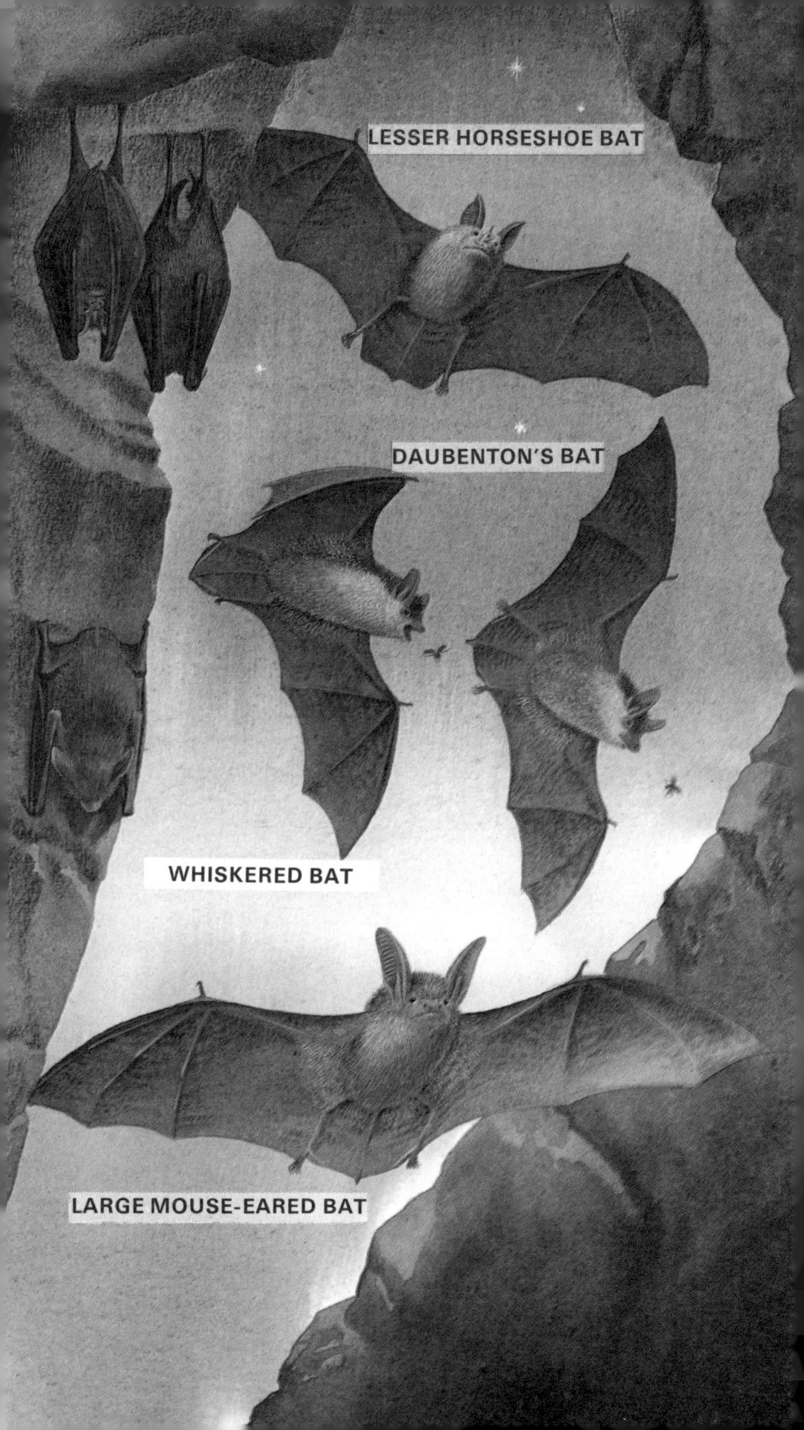

LESSER HORSESHOE BAT

DAUBENTON'S BAT

WHISKERED BAT

LARGE MOUSE-EARED BAT

PIPISTRELLE

PIPISTRELLE *(Pipistrellus pipistrellus).* This is the smallest as well as the most abundant and widespread of European bats. In early summer colonies composed entirely of females often take up residence in a crevice in the roof or walls of a house in order to have their young. Each bat produces a single baby which is left behind in the nursery when the mother leaves to feed. Within a month the young can fly strongly and the colony usually disperses. Length 7cm.

NOCTULE

NOCTULE *(Nyctalus noctula).* Large, narrow-winged bats that tend to fly high, taking over at dusk from the swifts in the pursuit of high-flying insects, noctules roost mainly in hollow trees and are rarely found in caves. Length 12cm.

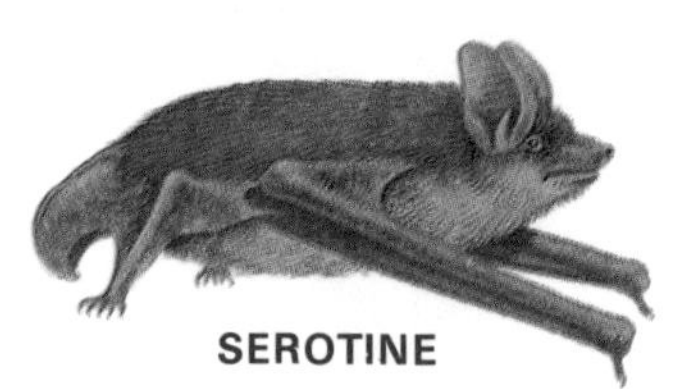

SEROTINE

SEROTINE *(Eptesicus serotinus).* Serotines are as large as noctules but have broad wings and a rather fluttering flight. Length 12cm.

BARBASTELLE

BARBASTELLE *(Barbastella barbastellus).* At close range barbastelles can be recognised by the strange black wrinkled face and the broad ears which meet on top of the head. They are very elusive creatures, roosting singly behind the bark of trees or in other inaccessible crevices. Length 10cm.

COMMON LONG-EARED BAT

COMMON LONG-EARED BAT *(Plecotus auritus).* The ears of a long-eared bat are so long (almost as long as its body) that when the animal is at rest they are folded concertina-wise and tucked under the wings. These bats tend to fly low, fluttering around the tree-tops and sometimes picking insects from the foliage. Length 9cm.

PIPISTRELLE
SEROTINE
NOCTULE
BARBASTELLE
COMMON LONG-EARED BAT
DWO
18.6.74

RABBIT *(Oryctolagus cuniculus)*. From their original home in Spain and Portugal rabbits have spread through most of Europe. In the Middle Ages they were preserved in warrens as a source of food; now they are considered agricultural pests. Litters of up to seven young can be produced every four weeks and the young can start breeding at an age of three or four months. On the other hand when the population density is high the birth rate is controlled by the unborn young being reabsorbed by the mother's body. Length 45cm.

MOUNTAIN HARE *(Lepus timidus)*. In most of their range these hares turn more or less completely white in winter and revert to a greyish brown coat in spring. However in the high arctic they may remain white all year whilst at the other extreme, in Ireland, they remain brown all year. In this way they maintain the maximum degree of camouflage to protect them against predators such as eagles, buzzards and foxes. They are hardy, surviving the arctic winter with little shelter, feeding on heathers and the bark of shrubs. Length 55cm.

BROWN HARE *(Lepus capensis)*. These are the lowland hares of Europe, renowned for their speed – they can reach 70km.p.h. Hares do not burrow and they have small litters, usually of two or three young. They are well developed at birth and can run within a few days. Hares are especially conspicuous in spring when they can be seen chasing each other in the fields. They indulge in curious 'boxing matches', perhaps to establish the right to territory. The sexes cannot be distinguished without close examination and it is therefore difficult to interpret their behaviour. Length 65cm.

RED SQUIRREL *(Sciurus vulgaris).* This attractive and lively animal is the only tree squirrel to be found in most parts of Europe, although in Britain red squirrels have suffered from competition with the introduced grey squirrels and they have become extinct in most of England during the last 50 years. They are only common now in the coniferous forests of Scotland. In Britain they are consistently a bright reddish brown but in many parts of continental Europe a dark greyish brown or almost black form can also be found. The prominent ear-tufts, which are so characteristic, are present only in winter. Red squirrels are most abundant in coniferous forests where they feed mainly on conifer seeds extracted by gnawing the scales off the cones. It can take as many as 200 cones to provide a reasonable day's food so they must work very hard for their living. Squirrels' nests, known as dreys, are usually built high in trees. They do not hibernate but in winter they are less active and spend much of their time in their nests. Length 42cm.

◁ **GREY SQUIRREL** *(Sciurus carolinensis).* This is the common squirrel of the deciduous woodlands of eastern North America. Introductions were made in various parts of Britain beginning about 1876 and they have been spreading ever since at the expense of the native red squirrel. The actual way in which this replacement takes place is not well understood but it probably involves competition for food during the most critical time of year, in late winter. Acorns are a favourite food and in the autumn much time is spent burying them, apparently at random, for consumption during the winter. Length 46cm.

EUROPEAN SUSLIK *(Spermophilus citellus).* ▷ Susliks are ground-living squirrels that are especially characteristic of the steppes of Russia and central Asia. This species extends westwards in Europe as far as Austria. Susliks live in colonies in open grassland and make deep burrows. In autumn they store great quantities of seeds and about September they retire underground for the winter. At first they feed on their stores of food but as winter approaches they become torpid and sleep until spring. Length 26cm.

FAT DORMOUSE
HAZEL
DORMOUSE
GARDEN DORMOUSE

FAT DORMOUSE *(Glis glis).* These elusive nocturnal dormice spend most of their time in the treetops in woods and orchards, feeding on buds, leaves, nuts and fruit. They become very fat in autumn, sometimes entering farmhouses to raid apple stores. They sleep throughout the winter in a nest in a tree-hole or in a crevice in a wall, occasionally in a building. Their alternative name, edible dormouse, comes from the ancient Roman custom of fattening them for the table. They were kept in special jars called 'gliraria.' They are widespread in most of central and southern Europe but in Britain they have been introduced and are confined to the Chiltern Hills. Length 30cm.

HAZEL DORMOUSE *(Muscardinus avellanarius).* Hazel dormice can be found in deciduous woodland wherever there is a dense tangle of shrubs near the ground. They climb about at night amongst the slender branches, feeding especially on buds and fruit. Usually, they are found asleep, tightly curled up in their winter nest, in a crevice at the base of a tree or amongst leaves under a bush. Many have been found hibernating in nest-boxes put up for birds. Length 15cm.

GARDEN DORMOUSE *(Eliomys quercinus).* These beautiful dormice spend more time on the ground than the others and are often found on rocky hillsides as well as in woods and gardens. Their diet is also more varied and includes insects and snails as well as fruit and nuts. They make a variety of whistling and churring noises. Like other dormice they hibernate throughout the winter, from October, or even as early as September, until April. Length 26cm.

NORTHERN BIRCH MOUSE *(Sicista betulina).* Although superficially like the much more common wood mice, birch mice are more like hazel dormice in behaviour. They live in woodland with dense shrubs and hibernate for the entire winter. They are unusual among rodents in feeding mainly on insects. This particular species is found in some parts of Scandinavia and northeastern Europe. A close relative is found in the steppes of southeastern Europe. Length 16cm. ▷

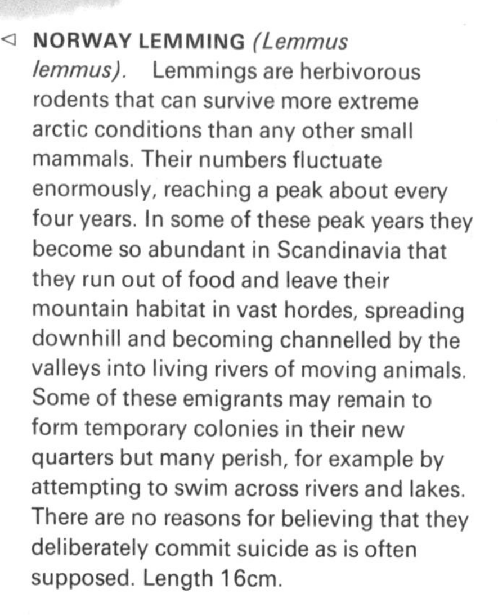

◁ **NORWAY LEMMING** *(Lemmus lemmus).* Lemmings are herbivorous rodents that can survive more extreme arctic conditions than any other small mammals. Their numbers fluctuate enormously, reaching a peak about every four years. In some of these peak years they become so abundant in Scandinavia that they run out of food and leave their mountain habitat in vast hordes, spreading downhill and becoming channelled by the valleys into living rivers of moving animals. Some of these emigrants may remain to form temporary colonies in their new quarters but many perish, for example by attempting to swim across rivers and lakes. There are no reasons for believing that they deliberately commit suicide as is often supposed. Length 16cm.

HAMSTER *(Cricetus cricetus).* The common hamsters of Europe are considerably larger than the familiar golden hamsters which are kept as pets (an Asiatic species), but like them have enormous expansible cheek-pouches in which they carry seeds to their underground stores. Hamsters are animals of open grassland but this species has adapted well to the margins of cultivated land and can be considered a minor agricultural pest. Like the superficially similar ground squirrels they hibernate throughout the winter. Length 30cm.

▽

GOLDEN HAMSTER

Δ

ALPINE MARMOT *(Marmota marmota).* Marmots are really very large ground squirrels. They live in colonies on open sunny mountain slopes above the tree line, making deep burrows if the depth of soil allows or occupying crevices under rocks, boulders or stabilized scree. Being active by day, they are familiar animals in the Alps and have even become picnic parasites' at some holiday resorts. Under more natural conditions they are very wary. An animal spotting danger utters a sharp whistle which sends the whole colony scuttling back to the shelter of the burrows. Marmots feed on a variety of green vegetation. They do not store food like the seed-eating ground squirrels, but like them they hibernate, sometimes a whole family curling up together for their winter sleep in a grass-lined nest deep down in a burrow. Length 70cm.

Δ

BEAVER *(Castor fiber).* Beavers are now rare animals in Europe, having suffered from centuries of exploitation for their fur as well as from loss of habitat by deforestation. The enormous flat scaly tail is a unique adaptation for their aquatic life and they are superb swimmers. In summer they feed upon a great variety of green vegetation, but their unique specialization is the way they provide for their winter sustenance. They fell young hardwood trees, using their enormous orange incisor teeth, float the branches to a pond and store them under water to provide a constant supply of succulent bark throughout the winter.

Beavers may live in a simple hole in a river bank, but they sometimes construct much more elaborate living quarters. A lodge, built in a pool, consists of a pile of branches with a dry platform above water level and a roof of twigs and mud. The entrance to the chamber is under water. To be effective this requires a constant water level and this is achieved by building a dam across the stream, using branches and clods of earth or turf. Length 115cm.

MUSKRAT

MUSKRAT *(Ondatra zibethicus).* In many parts of Europe muskrats have escaped from fur farms and become established under natural conditions. Really large aquatic voles native to North America and known in the fur trade as musquash, they feed on riverside vegetation and make burrows in the banks of rivers and canals, causing considerable damage in areas that are liable to flooding, for example in Holland. They became established in Britain in the 1930s but were successfully eradicated. Length 60cm.

COYPU *(Myocastor coypus).* Coypu are very large aquatic rodents related to guinea pigs and like them native to South America. They are bred for their fur and are known as nutria in the fur trade. Escaped animals have become well established in waterways of eastern England. There are also colonies on the continent but they survive less well there since they are very vulnerable in severe winters. Length 90cm.

▽

BANK VOLE *(Clethrionomys glareolus).* These voles can be found wherever there are shrubs, for example in hedgerows or amongst the undergrowth of brambles in woodland. They feed on leaves and fruit and being diurnal are not too difficult to watch as they run nimbly amongst the branches or shrubs or fallen trees. Length 14cm.

GREY-SIDED VOLE *(Clethrionomys rufocanus).* A northern relative of the bank vole, living amongst heather and other dwarf shrubs on the mountains and tundra of Scandinavia. Like other northern rodents they undergo cycles of abundance and scarcity but without reaching such plague densities as do the lemmings. Length 15cm.

COMMON PINE VOLE *(Pitymys subterraneus).* This is the most subterranean of the European voles, making extensive tunnels just under the surface in grassland or open woodland. Like the other voles they are herbivores, adapted to feeding on very tough vegetation like grass and roots. In keeping with their underground life the eyes and ears are particularly small. (Several closely similar species occur in southern Europe and in the mountains of central Europe.) Length 12cm.

COMMON VOLE *(Microtus arvalis).* These voles are rather less subterranean than the pine voles but can nevertheless survive on quite closely cropped pasture, making a network of runways just at ground level. They are common in much of Europe, although confined to the Orkneys and Channel Islands in Britain, and can cause considerable damage to crops when the numbers are high. Length 13cm.

FIELD VOLE *(Microtus agrestis).* Like the common vole this is a grassland animal but it is less of a digger and likes longer grass to protect it from its many enemies. Voles of all kinds breed prolifically and are important items in the diet of many predators such as owls, kestrels, harriers, weasels, cats, foxes and snakes. Length 14cm.

WATER VOLE *(Arvicola terrestris).* In many areas these large voles are aquatic, as their name implies, living in burrows on the banks of ponds and slow rivers wherever there is dense vegetation to provide food and cover. They have no obvious adaptations for an aquatic life. They swim by paddling along with all four feet and show little of the speed and agility of the more specialized aquatic mammals like the otters, beavers or even their close relatives the musk rats. In other areas, for example in the Alps, they are much less aquatic, living a rather subterranean life in grassland and throwing up hills of earth that could be mistaken for mole hills. Length 30cm.

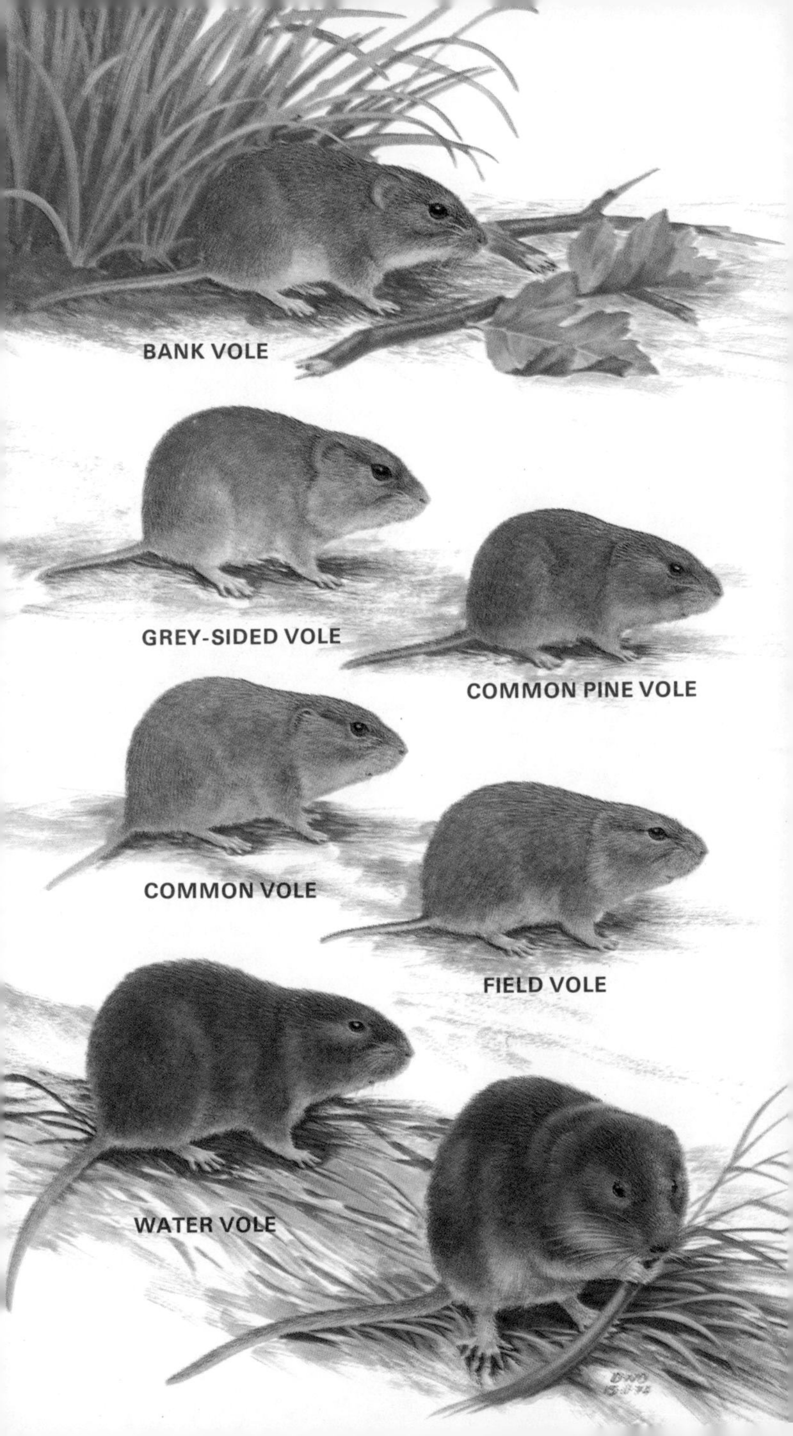
BANK VOLE
GREY-SIDED VOLE
COMMON PINE VOLE
COMMON VOLE
FIELD VOLE
WATER VOLE

◁ **HARVEST MOUSE** *(Micromys minutus).* The intricate spherical nests woven from shredded grass by these mice are easier to find than the animals themselves. These nests are occupied in summer when the animals spend much of their time above the ground, climbing with great agility amongst the stems of grasses and other tall herbaceous plants to feed on seeds. Although harvest mice will live in fields of cereal they also occur commonly in hedgerows and on the edges of woods. Length 13cm.

STRIPED FIELD MOUSE *(Apodemus agrarius).* The dark stripe of this eastern species of mouse makes it easy to distinguish from its close relatives but gives it a superficial resemblance to the distantly related birch mice (p. 29). Length 17cm.

WOOD MOUSE *(Apodemus sylvaticus).* This is the most widespread, abundant and versatile species of mouse in Europe, although, being strictly nocturnal, it is rarely seen. Wood mice live in a great variety of habitats from dense woodland to hedgerows and gardens, and often enter buildings in search of food. Length 20cm.

YELLOW-NECKED MOUSE *(Apodemus flavicollis).* These mice are very similar to wood mice. They are extremely agile and move about freely in the highest canopy of forest trees. They are even more inclined than wood mice to enter houses, being especially attracted by stored apples. Length 22cm.

HOUSE MOUSE *(Mus musculus).* House mice are found not only in buildings but also in farmyards, gardens, and hedgerows and in cultivated crops. These outdoor animals may be very pale, especially in southern and eastern Europe and can easily be mistaken for the other outdoor species of mice. (*See also next page.*)

HARVEST MOUSE
STRIPED FIELD MOUSE
WOOD MOUSE
YELLOW-NECKED MOUSE
HOUSE MOUSE

Wild type

HOUSE MOUSE *(Mus musculus).* Wherever there is human activity there are house mice, even in such unlikely places as coal mines and cold storage warehouses. In favourable conditions they breed prolifically – many hundreds may live in a single corn rick. They are exceedingly destructive, spoiling much more food than they actually consume and transmitting diseases. The great variety of 'fancy' mice kept as pets are descended from the House mouse, as are the white mice used in biological and medical laboratories. Length 17cm. (*See also previous page.*)

Fancy breeds

SHIP RAT *(Rattus rattus).* In all the warmer parts of the world, including the Mediterranean coast of Europe, the ship rat is ubiquitous. In northern Europe they are confined to ports where they have escaped from the holds of ships. They are agile climbers, just as likely to be found in attics as in cellars. Although often called the black rat, ship rats may be black or brown, but they can be distinguished from common rats by their longer ears and tails. Length 40cm.

HOUSE MICE

COMMON RAT *(Rattus norvegicus).* This is the dominant rat in most of Europe. They are extremely versatile animals, living in a great variety of habitats from city sewers and warehouses to farmyards, cereal crops, rocky shores and salt marshes. They breed prolifically – litters of up to about twelve young can be produced at intervals of three or four weeks. Grain is their favourite food but they can survive on such improbable substitutes as candles and soap.

The white rats used as laboratory animals belong to this species. They share the fast breeding potential and rapid development of the wild animals but have been bred for docility, being placid and easy to handle in contrast to their nervous and excitable wild relatives. Length 44cm.

Albino

COMMON RAT

SHIP RAT
COMMON RAT
Dark form

BROWN BEAR *(Ursus arctos).* Bears once occupied the whole of Europe, including Britain, but have been reduced to a few small remnant populations in secluded forests mainly in mountain regions. They have a reputation for ferocity and may fight back if they are persecuted, but if left alone they are peaceful animals – unprovoked attacks on people are almost unknown. Their love of honey is well known but this is just one of a great range of items that feature in their diet. They can be described as opportunists, making use of whatever food source happens to be locally or temporarily abundant – fish, rabbits, apples, acorns. They become very fat in autumn (raiding crops if natural food is scarce) and retire to secluded dens for the winter. The young are born in the winter dens and are remarkably small at birth, no bigger than a guinea pig. Length 200cm.

WOLF *(Canis lupus).* Wolves are commonly associated in people's minds with dense forest but they have only been driven there by persecution. They are basically animals of open country where they can deploy to the full their tactics of hunting in packs for large prey, usually deer. Hunting is usually done by day and both visual and sound signals are used to coordinate the pack as they attempt to outrun and outwit their prey. However, in Europe they now tend to be more furtive, nocturnal hunters. The wolf is the ancestor of the domestic dog and many of the familiar habits of the dog, such as tail wagging and submission to a master, have their origins in the social organization of the wolf pack. Wolves of course can be destructive of domestic livestock and very few remain in western Europe. Length 150cm.

ARCTIC FOX

Summer coat

Autumn coat

Winter coat

Blue form

◁ **RED FOX** *(Vulpes vulpes).* Within the dog family foxes and wolves are, in many ways, at opposite ends of the spectrum. Foxes are small, they hunt alone rather than in packs, and they are adapted to feeding at night on small prey. These factors, along with their renowned cunning and versatility, have enabled red foxes to adapt well to man-made changes in the environment, even in cities and industrial areas. Early morning commuters in London frequently see foxes on railway embankments or factory sites quite close to the centre of the city. Foxes feed mainly on rodents and a den of hungry cubs – there may be as many as ten in a litter although five is more usual – can consume an enormous quantity of rats or voles. Length 110cm.

ARCTIC FOX *(Alopex lagopus).* Like ▷ several other mammals and birds living all year on the arctic tundra the majority of arctic foxes moult into a pure white coat as winter approaches. But a small proportion of animals, known as blue foxes, remain a smoky grey colour all year. In summer arctic foxes have an abundance of food in the form of nesting birds and their eggs as well as voles and lemmings. In winter most of the birds leave and, if rodents are scarce, the foxes may be driven to feeding on shellfish on the coasts or else to migrate southwards to the edge of the forests. Length 90cm.

◁ **BADGER** *(Meles meles).* Badgers are strictly nocturnal which makes them difficult to sight, but their presence is usually easily detected by their elaborate systems of holes and tunnels, known as sets. These are commonly tunnelled in a bank, at the edge of a wood for example, and open out at several holes, each of which has a large pile of earth in front of it. The animals normally spend the winter less actively, though not in hibernation, and when spring arrives, they will throw out the old bedding from their nests, set deep underground at the end of a tunnel, to replace it with fresh grass and leaves ready for the birth of their young. The cubs are usually born two or three at a time, and do not emerge from the set until half-grown, when they may sometimes be seen rolling about and chasing each other in front of the hole. Badgers feed largely on earthworms, but are opportunists in the same way as bears, and their diet may include a wide variety of animal and vegetable food. Length 85cm.

RACCOON *(Procyon lotor).* Lively, inquisitive and versatile, raccoons have adapted well to the dramatic changes inflicted on the countryside in their native North America. They may therefore be able to spread successfully in Europe, where they first became established in the wild as a result of escapes from captivity in West Germany. Their diet is mainly vegetarian, but includes a wide variety of animal food – eggs, frogs, fish, crayfish, rodents. Length 80cm.

RACCOON DOG *(Nyctereutes procyonoides).* This true dog is not closely related to the raccoon. Native to eastern Asia, and introduced in Russia it has spread across central Europe. Length 75cm. ▷

RACCOON

◁ **STOAT** *(Mustela erminea).* The stoat can always be distinguished by its black tail-tip whether it is in its brown summer or white winter coat. Throughout most of northern Europe all animals turn white in winter, when they are known as ermine, but there is a zone that runs from Britain across central Europe where the change may be only partial, varying in extent from animal to animal and from year to year. Stoats are fierce little predators whose quarry consists mainly of mice and voles; but it is not unusual for them to kill even larger creatures. A stoat is capable of mesmerizing a rabbit, for example, by dancing around and dodging in and out of cover until it finally dashes in to the kill with a bite on the neck. Length 35cm.

◁ **WEASEL** *(Mustela nivalis).* Weasels are even more specialised mouse- and vole-hunters than stoats, being small enough to pursue them to the extremities of their own runs and tunnels. Since they spend much of their life underground or in dense cover, it is only those inhabiting the extreme north of Europe that need and attain a white winter coat. But in spring particularly, when they add nesting birds to their diet, weasels may be seen running with great confidence and agility along the slender twigs of shrubs and trees. Length 27cm.

POLECAT *(Mustela putorius).* ▷
More strictly nocturnal than their smaller relatives, the stoats and weasels, polecats will live on farmland if not persecuted, and in woodland, having a predilection for wet places where they can find the frogs that form a large part of their diet along with rodents, rabbits and birds. They have scent glands under the base of the tail, used for marking territory, a feature common to all members of the weasel family, but in the polecat the secretion they produce is particularly pungent and probably also serves to deter predators. The dark species found in western Europe is replaced by a paler species in eastern Europe, the **STEPPE POLECAT** *(M. eversmanni).* Length 55cm.

POLECAT

STEPPE POLECAT

FERRET This domesticated form of ▷
the wild polecat, used for hunting rabbits, is probably descended from the Steppe polecat. Many are a uniform creamy white, but cross-breeds, appropriately called 'polecat-ferrets,' show a dilute form of the colour and pattern of their wild parent. In some areas where true wild polecats have become extinct, Scotland for example, escaped ferrets have established populations in the wild, among which brown forms tend to predominate.

△

MINK *(Mustela lutreola).* Renowned for their beautiful, sleek fur, mink are semi-aquatic weasels that usually live by forest streams. They swim and dive expertly in pursuit of their prey – most often water voles but also frogs and water birds. The European mink, that can be recognized by the white on the upper lip, is now a rare animal in western Europe; whereas the similar American mink *(Mustela vison)*, the one usually farmed for its fur, has escaped to become more abundant than the native species in many parts of Europe including Britain. Length 50cm.

OTTER *(Lutra lutra).* Otters are fast and powerful swimmers, propelling themselves through the water by sinuous movements of the whole body and tail similar to a fish's, although they flex up and down rather than from side to side, and make use of their legs to steer and paddle at low speeds. They are thus well adapted for hunting their main diet of fish in their own element, while also feeding on crabs in coastal waters; but their vulnerability to pollution and disturbance has made them extinct on many lowland rivers in industrial areas. Young otters are very playful and delight particularly in sliding down a muddy bank into the water. Length 120cm. ▷

BEECH MARTEN

◁ **PINE MARTEN** *(Martes martes).* More arboreal than any other members of the weasel family, pine martens are most at home in the coniferous forests their name implies. They race through the tree-tops and leap prodigiously from branch to branch in pursuit of squirrels and birds with the same ease that other animals display over ground. But they are adaptable and can live equally well on rocky hillsides and cliff areas far from any trees, such as in Ireland where other carnivores are scarce. Pine martens have long been trapped for their fur in the northern forests and heavily persecuted elsewhere in the interests of game preservation. Nevertheless, they have survived moderately well and are beginning to spread and increase in places, such as Scotland, where combined persecution and deforestation greatly restricted their range in the past. Length 75cm.

BEECH MARTEN *(Martes foina).* Very similar to pine martens, these animals differ in appearance by the throat patch which is pure white without any trace of yellow. Having a more southern range, they hunt both on the ground and in trees in the deciduous forest they inhabit, and will eat voles, mice, rabbits, frogs and lizards as well as birds. Nocturnal like pine martens, and more strictly so when, as happens frequently, they live near farms and villages, they make their dens amongst rocks, in a hollow tree, or even in a derelict building. Their young are born in spring, normally in a litter of four or five. Length 70cm.

△

WOLVERINE *(Gulo gulo).* The most heavily-built of the weasel family, these bear-like animals live in the northern forests and hunt by stealth, often lying in wait on a tree-branch until they drop on to an unsuspecting grouse or even small deer. Also known as 'gluttons,' they can gorge themselves on fiercely defended carrion in winter to guard against possible hunger in ensuing days, and are very unpopular with fur trappers, whose catches they often depredate. Length 90cm.

◁ **WILD CAT** *(Felis silvestris).* Ancestor of the domestic cat, this species has interbred to a small extent with the feral domestic cat in most of Europe, although not enough to affect the true wild cat's appearance, characterised by its large size and bushy tail. Contrary to their reputation, wild cats are not normally vicious creatures, but rather elusive, rarely attacking anything larger than a rabbit or crow. They are more at home in trees than the average tabby but do much of their hunting on the ground, stalking prey of mice, voles, frogs and small birds until they are in a position for a lightning pounce. Length 100cm.

PARDEL LYNX *(Felis lynx pardina).* This southern form of the lynx is now rare although strictly protected in certain areas where it survives, such as the Coto Doñana of south-western Spain. It lies in wait to pounce on its prey of a rabbit or hare, grasping it in powerful paws. Length 110cm.

NORTHERN LYNX *(Felis lynx lynx).* Once also found in the Alps, this is the only member of the cat family in the northern coniferous forests of Scandinavia. Lynxes hunt by stealth, like most cats, often keeping vigil from a low branch until they drop silently on to prey. Hares and grouse are their main victims and fur-trappers' records show that the numbers of hares and of lynxes tend to fluctuate in unison. Kittens, born in summer in litters of two or three, spend their first difficult, northern winter with the mother, learning to hunt, and only become independent ◁ when spring brings more easily caught food. Length 130cm.

GENET *(Genetta genetta).* These wonderfully lithe and agile relatives of the mongoose and civet are mainly African. The species found in south-western Europe is one of the most widespread in Africa and was possibly introduced to Europe by human agency. Length 100cm.

MONK SEAL *(Monachus monachus).* This is the only kind of seal in the Mediterranean, surviving in very small breeding colonies, possibly in Corsica and amongst the Yugoslavian and Greek islands. Persecution by fishermen and disturbance of their breeding beaches by holidaymakers make it seriously endangered. Length 250cm.

RINGED SEAL *(Phoca hispida).* This is the commonest seal of arctic coasts, also found in the Baltic and some adjacent freshwater lakes. The young are born on land-fast ice in the spring, where they remain for almost two months before taking to the water. Length 130cm.

COMMON SEAL *(Phoca vitulina).* The commonest seal in most of northwestern Europe, found especially on sheltered coasts and in sandy bays and estuaries. They breed in early summer and usually give birth on sandbanks during low tide to pups that are ready to take to the water as the tide rises. As with other seals, the diet consists mainly of fish but a certain amount of shellfish may also be taken. Length 160cm.

GREY SEAL *(Halichoerus grypus).* Found mainly on more exposed coasts than the common seal, grey seals often pull themselves out on to rocky ledges during low tide, where their motley coats blend well with the rocks and seaweed. They almost invariably breed on the beaches of small rocky islands in late autumn, tending to congregate in a relatively small number of large colonies. The pups remain on land for about three weeks after birth, when they are deserted by their mothers and, shedding their original white silky coats, make their own way down to the sea. During this time the older males (bulls) hold territories on the beach and mate with a number of females. The situation is different in the Baltic, however, where the young are born in small colonies on the ice in early spring. The European population numbers about 50,000 animals, of which approximately 40,000 are around the British Isles. Length 300cm.

COMMON SEAL
♀
GREY SEAL
♂

Key to Deer, Sheep and Goats

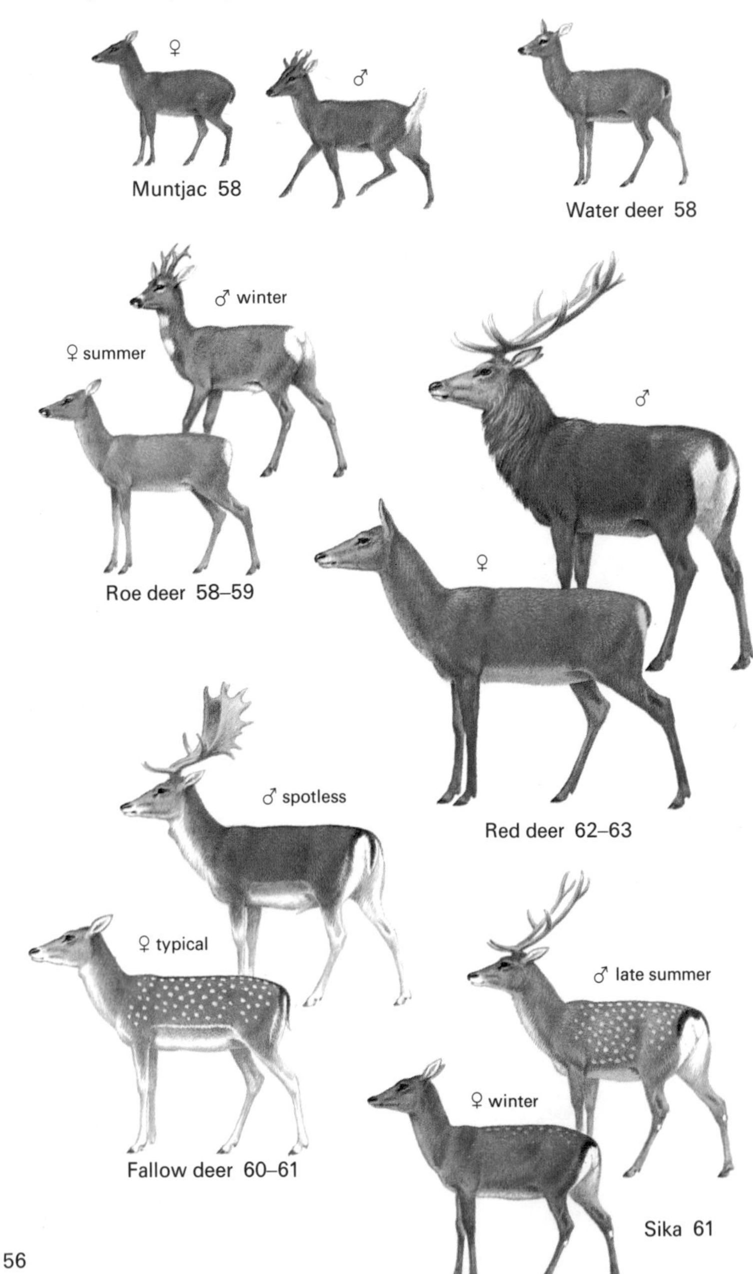

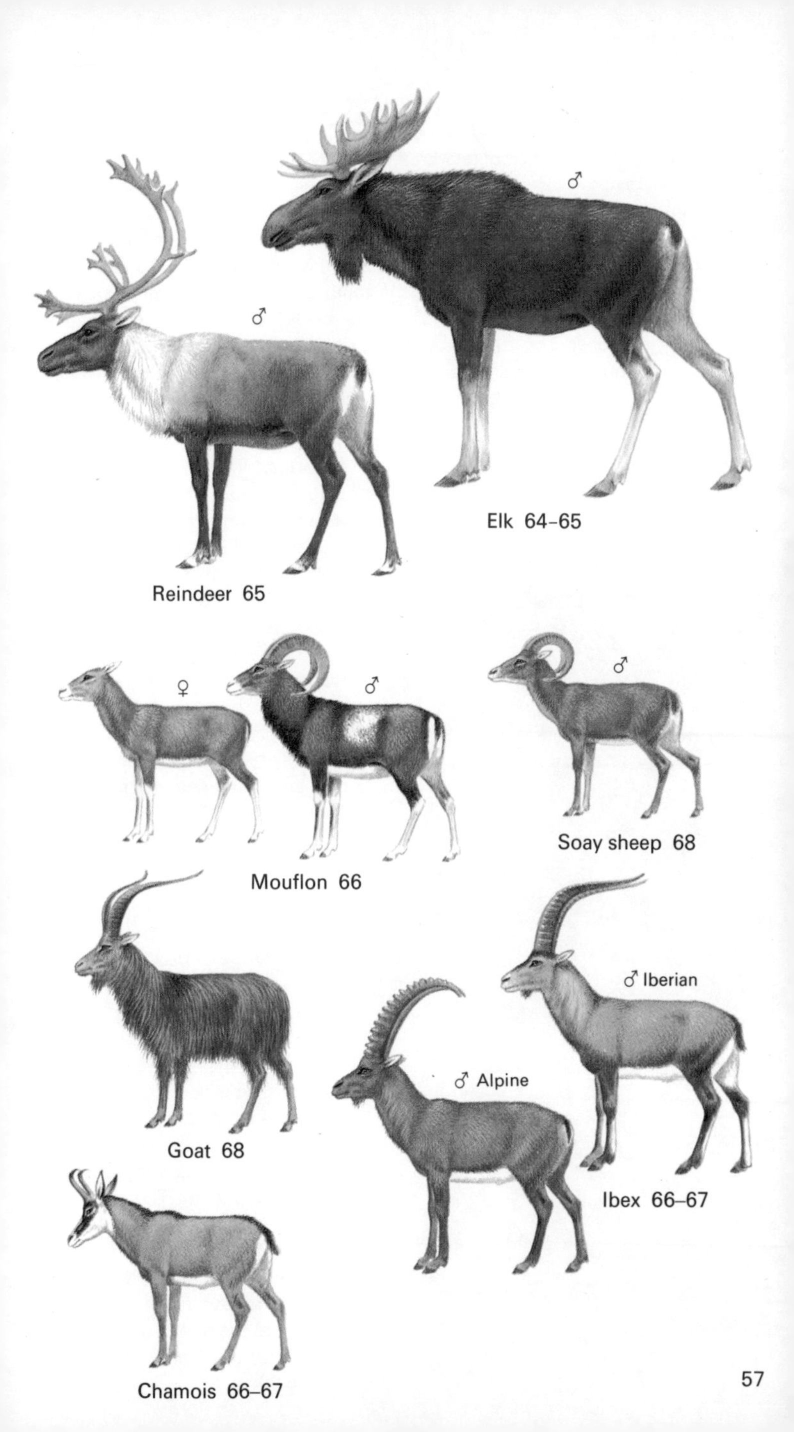
♂
♂
Elk 64–65
Reindeer 65
♀
♂
♂
Soay sheep 68
Mouflon 66
♂ Iberian
♂ Alpine
Goat 68
Ibex 66–67
Chamois 66–67

ROE DEER *(Capreolus capreolus).* The roe is the smallest and most widespread of the native European deer. They are generally found alone or in pairs, being more solitary creatures than their larger relatives and will bound away if disturbed in woodland, showing up the white rump patch that is particularly conspicuous in winter. The winter coat differs markedly from the summer, changing in autumn from a bright orange-brown to a dull greyish-brown; while the antlers of the males (bucks), used to strip bark from sapling trees during the mating season in summer, are fully grown by that time and shed in late autumn, which is much earlier than in the larger deer. The 'frayed' trees that result from this habit of theirs are a useful sign that roe are present. They feed mainly by browsing on leaves of small trees and shrubs but they also graze and can be sighted most easily in early morning when they emerge from the woods to feed on adjacent grassland. Length 110cm.

MUNTJAC *(Muntiacus reevesi).* Originally from China, this little deer has been introduced in England and France. They are difficult to sight as they live in dense woodland, but their presence can be detected by their loud bark – a single sharp cry repeated every few seconds. The young may be born at any time of year and for the first few days are carefully hidden in a thicket until they are strong enough to follow their mothers. Muntjac browse on the leaves of bramble and other woodland shrubs. Length 85cm.

CHINESE WATER DEER *(Hydropotes inermis).* Like the muntjac this introduction from China has become established in parts of England and France. They are remarkable amongst deer for their lack of antlers, instead of which the males have enlarged canine teeth in the form of long, slender tusks that grow from the upper jaw. They prefer to live in woodland, particularly in the vicinity of marshes and reed-beds, and usually will bear from two to four in a litter, in contrast to the one or occasional two typical in other kinds of deer. Length 90cm.

Winter
♂
Summer
♀
♂

Winter ♂
Summer ♂
Young ♂
♀

SIKA DEER *(Cervus nippon).* Originally from eastern Asia, it is mainly the Japanese sika that has been introduced in Europe. They resemble the larger native red deer and the fallow in both appearance and behaviour but during the rutting season in autumn the males emit a penetrating whistle very different from the red stag's roar. There is interbreeding between sika and the indigenous red deer in some areas, such as northwestern England. But sika can be distinguished reliably from fallow by the tail, which is shorter and wholly white irrespective of sex, age or season. Length 120cm.

◁ **FALLOW DEER** *(Cervus dama).* These probably originated in the Mediterranean area but are now the commonest deer in the parks and woodlands of much of western Europe. Though most often spotted boldly in summer and faintly in winter, there are various colour forms: a very pale one retains the same spots at all seasons while a common dark form is consistently unspotted. When rutting in autumn the older males (bucks) join the herds of females and young animals, and, making a rhythmic grunting noise, they fight by pushing with their antlers. They fast during the rut but, with a good crop of acorns, they can quickly regain their condition against the rigours of winter. When the mothers are about to give birth, in May or June, they separate from the herd and produce their fawns in dense vegetation such as bracken where they hide them until strong enough to run with the herd. Length 150cm.

Stag
Young stag

Hind and fawn

RED DEER *(Cervus elaphus).* These are the largest deer in most of western Europe. Although basically woodland animals, in some areas such as the Highlands of Scotland they inhabit open hill country and move down to low ground only in severe winter weather. For most of the year the males (stags) live in herds apart from the females (hinds) and young. But in autumn during the time known as the 'rut,' the stags try individually to attach themselves to a group of hinds, roaring, wallowing in mud and fighting with antlers interlocked. Although the strongest will remain attached to the herd, damage to the loser is usually more psychological than physical, and he may well return to usurp his rival later in the rut. In spring the antlers are shed. New ones grow during summer beneath a velvety skin that is discarded when they reach full size ready for the next rutting season. Length 250cm.

△

REINDEER *(Rangifer tarandus).* These, the characteristic deer of the far north, once roamed the summer tundra in great herds, migrating southwards to the Scandinavian woodlands in winter. But now the true wild animals survive only sparsely in Scandinavia, while in Lapland domesticated reindeer provide food, clothing, shelter and transport for the nomadic Lapps. Their wide hooves enable them to walk on snow and dig beneath it to reach their winter diet of lichens. Length 210cm.

◁ **ELK** *(Alces alces).* Elk are the largest of all living deer – an old bull stands considerably taller than a man. They lead solitary lives in the northern coniferous forests, often by rivers and lakes where, during summer, they may be seen to wade deeply or even swim as they feed on succulent waterside vegetation. In Russia attempts at domestication are succeeding. Length 260cm.

Δ

MOUFLON *(Ovis musimon).* The wild ancestors of our domestic sheep live mainly in the mountains of Asia but this one race, the mouflon, is found in Cyprus, Sardinia and Corsica and has been introduced from there to many parts of continental Europe. They inhabit the forest and open slopes on high ground, or dense woodland from which they will emerge to graze in clearings at dusk. The mouflon's woolly undercoat, original of the domestic sheep's fleece, is concealed by a coat of straight hair. Length 120cm. (*See next page for feral domestic sheep.*)

IBEX *(Capra ibex).* Tough and sure-footed, these true wild mountain goats live above the tree-line of the Alps, Pyrenees and other Spanish mountain ranges even in winter, spending the day at a higher altitude than at night when they descend to graze the alpine meadows. Strict protection has now secured the flourishing of the Alpine ibex, once almost hunted to extinction; but the Spanish ibex, with lyre-like horns curving forward at the tips, is still in a precarious position. Length 150cm. (*See next page for other wild goats.*)

CHAMOIS *(Rupicapra rupicapra).* As sure-footed as ibex, chamois tend to live close to the tree-line at lower altitudes, even moving into the shelter of the forests in winter. Groups of females and young, united in autumn into larger herds, are joined by males who, erecting a line of longer hair along the back to appear large and intimidating, indulge in a great deal of chasing, accompanied by deep rumbling bleats. Length 115cm.

♂
♀
IBEX
Winter
CHAMOIS
Summer

FERAL GOATS The wild goats *(Capra aegagrus)* of the eastern Mediterranean are the ancestors of the domestic goat and some, with horns curving backwards in a single plane and probably little affected by domestication, survive on a few small Greek islands and on Crete. But feral goats, which generally have a spiral twist to the horns, are derived from escaped domestic animals and live on many other small islands and in several parts of upland Britain. Goats are able to survive on the toughest of vegetation and to reach rocky ledges that are not safely accessible to sheep or deer. Mature males, or billies, live by themselves, joining the herds only in autumn during the rut. The kids are born in early spring, usually singly, and soon are as agile on stony ground as their parents.

SOAY SHEEP This primitive breed of domesticated sheep has survived unchanged for centuries on the island of Soay in the St Kilda group in Scotland, and still lives there and on the adjacent Hirta under natural conditions. But now they can also be seen in many European parks and zoos. They have the true woolly fleece of the domesticated sheep but otherwise greatly resemble the ancestral wild sheep – such as the mouflon. (*See previous page*.)

△

PRIMITIVE CATTLE Although the wild ox or aurochs became extinct by the 17th century, all European breeds are in fact derived from it. Some primitive breeds survive in Hungary and France, while in several British parks herds of white cattle have been maintained for many centuries without any attempts at improvement and without crossing with modern breeds. One of the best known and purest bred of these herds is at Chillingham in northeastern England where at any one time a single mature bull dominates the herd, ensuring a considerable degree of in-breeding. So that while no truly feral – that is uncontrolled – cattle exist in Europe, these herds do provide some clue as to the appearance of the domestic cattle of the middle ages and earlier.

PRIMITIVE HORSES The wild horse *(Equus)* that once ranged across the Eurasian steppes from eastern Europe to Mongolia, ancestor of all domestic breeds and still kept in a moderately pure-bred form in many parks and zoos, probably survives only sparsely in Sinkiang in its true eastern form, known as the Przewalski horse *(Equus ferus przewalskii)*. Attempts have been made to 'reconstruct' the western form, the tarpan *(E. f. ferus)*, extinct after 1850, by selective breeding of domesticated animals. In Poland a herd of tarpans lives in a semi-wild state, as do primitive breeds in other parts of Europe that retain their local traits because they are not crossbred or strongly selected. The erect mane, absence of forelock, pale muzzle and faint cross-striping on the legs are characteristics of the wild horse that many of their domestic descendants have lost. For no reason, except their smaller size in relation to other working horses, these are called 'ponies.' New Forest and Dartmoor are two typical semi-wild British breeds, while the Dülmen and Camargue are ancient breeds in Germany and the Rhone delta of southern France respectively.

Tarpan
Carmargue
Dülmen

△

EUROPEAN BISON *(Bison bonasus).* Herds of bison once roamed throughout western Europe, but the destruction of forests, their principal habitat, coupled with over-hunting gradually diminished them until in the 1920s the last two remaining herds, in Lithuania and the Caucasus, became extinct. The species survived only in zoos and parks and from these a herd was established in the Bialowieza Forest in Poland where they are now thriving. Bison live in small herds consisting of cows and young animals of both sexes, led by an experienced cow. Mature bulls live a more solitary life, joining the herds only in autumn. They stake their claim by churning the soil with their horns and scraping the bark from trees. Bison feed by browsing on the foliage of trees in summer and on heather and other dwarf shrubs in winter. Length 300cm.

WILD BOAR *(Sus scrofa).* The wild ▷ boar's reputation for ferocity is valid enough where wounded or cornered animals are concerned, but if not persecuted they are inoffensive and elusive creatures, well able to keep out of sight in the dense woodland they prefer. Although omnivorous, this ancestor of the domestic pig is mainly vegetarian, eating enormous quantities of acorns or beech mast in autumn to lay down a store of food, in the form of fat, for sustenance during the difficult days of winter. Up to 5kg of acorns have been found in the stomach of one animal. The young are usually born in spring in a litter of up to ten, and their striped coats allow them to pass unnoticed in the dappled light of the forest. Length 180cm.

FIRE SALAMANDER *(Salamandra salamandra).* Found in cool, moist, often forested areas over much of Europe, this slow-moving, nocturnal animal is usually active after rain. Its vivid colouring warns predators that it is inedible for the skin produces a poisonous secretion. Instead of laying eggs, females usually give birth to tadpoles that develop the characteristic adult colouring even before they leave the water. Length 20cm.

ALPINE SALAMANDER *(Salamandra atra).* Lives only high in the Alps and neighbouring mountains where it is so cold that the young, which are usually born in pairs, may take four years to develop inside the mother. Length 16cm.

ITALIAN CAVE SALAMANDER *(Hydromantes italicus).* Only rarely found outside moist caves, this salamander climbs on their walls catching its insect prey with its long, sticky tongue. Females may stay by their eggs for over a year. Apart from another species on Sardinia, its nearest relatives are in California. Length 12cm.

PYRENEAN BROOK SALAMANDER *(Euproctus asper).* A nocturnal salamander that lives in and around cold mountain lakes and streams. In spring, males grasp their mates with their prehensile tails. Length 16cm.

SPECTACLED SALAMANDER *(Salamandrina terdigitata).* So called because of the light mark joining the eyes, this western Italian salamander hunts mainly by night along the edges of rocky streams. If captured it may feign death or display the red underside of its tail. Length 10cm.

ALPINE SALAMANDER
ITALIAN CAVE SALAMANDER
PYRENEAN BROOK SALAMANDER
SPECTACLED SALAMANDER

ALPINE NEWT *(Triturus alpestris).* Most newts spend much of their time on land, leading a secretive life under logs and stones and amongst dense herbage and leaf litter, but in the spring they migrate to water for breeding when males become more brightly coloured and develop a flexible crest on the tail and often the back as well. They court the females in a complex ritual in which the tail may be alternately lashed towards their mate and vibrated close to their own body. Although many kinds of newts look rather similar on land, the breeding dresses of the males are very different and this probably helps the females pick out the right partner.

The Apine newt often lives in mountain areas, but in the north of its range it may be found in the lowlands, especially in rather shady pools. It spends more time in water than most newts. Length 12cm.

SMOOTH NEWT *(Triturus vulgaris).* Although not present in southwest Europe, this is one of the most widespread species of newts and one of the least aquatic outside the breeding season. In southeast Europe it can look very like the next species, but Smooth newts usually have a spotted throat whereas Palmate newts do not. Length 11cm.

PALMATE NEWT *(Triturus helveticus).* A western relative of the Smooth newt, it overlaps with this species in some areas including Britain. The two are often found in the same ponds but the Palmate newt predominates in more acid waters. Length 9cm.

WARTY NEWT *(Triturus cristatus).* One of the largest newts, it rarely seems to travel far from its breeding ponds and may sometimes remain in the water after the breeding season is over. Length 14cm.

MARBLED NEWT *(Triturus marmoratus).* The Marbled newt replaces the Warty newt in southwest Europe, but they overlap in parts of France where hybrids are sometimes found. It seems to prefer drier habitats on leaving the water, when it becomes a much more brilliant emerald green. Females and young have an orange stripe on the back. Length 14cm.

SHARP-RIBBED SALAMANDER *(Pleurodeles waltl).* This very big salamander lives in ponds and sluggish streams in Spain and Portugal and is often completely aquatic. It gets its name from its very pointed ribs which frequently stick through the orange warts along its sides. Length 30cm.

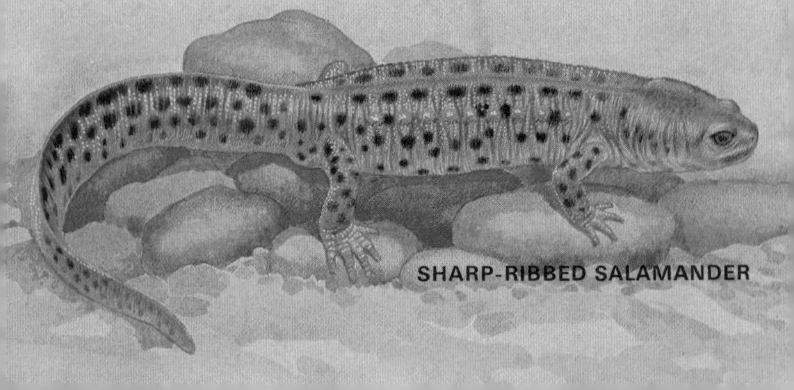

SHARP-RIBBED SALAMANDER

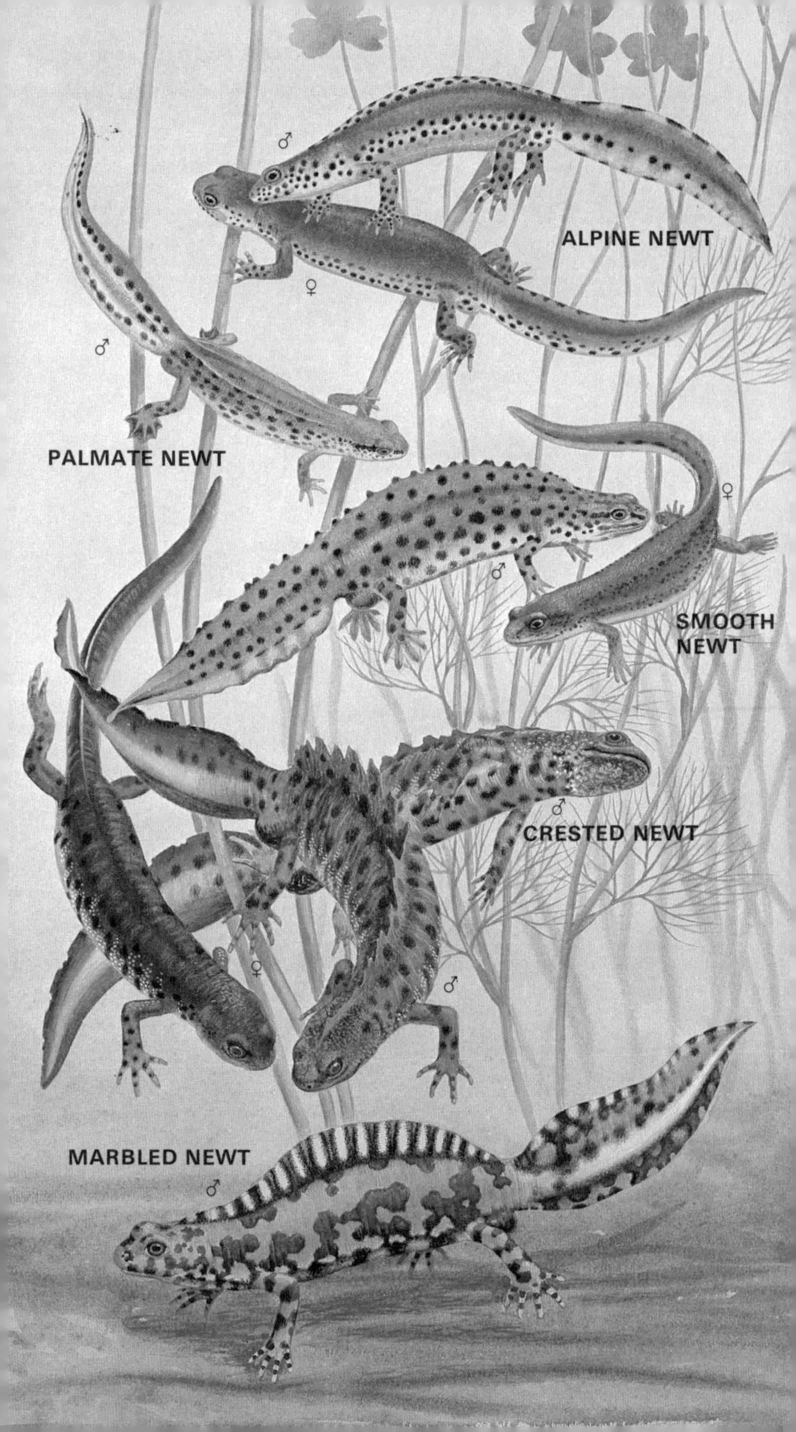
♂
ALPINE NEWT
♀
♂
PALMATE NEWT
♀
♂
SMOOTH
NEWT
♂
CRESTED NEWT
♀
♂
MARBLED NEWT
♂

EDIBLE FROG
POOL FROG
MARSH FROG
AMERICAN BULLFROG

GREEN FROGS Many European pools and rivers are enlivened in summer by hordes of often brightly coloured frogs that sit along the water's edge or float lazily on the surface, warming themselves in the sun. These are known as Green frogs and, although not always green, they can easily be recognised by the vocal sacs of the males which are blown out like a small balloon at each corner of the mouth when the frogs call. The sacs amplify the harsh, laughing croak which is very varied and can be heard day and night, being at its peak in the spring. Green frogs are enthusiastic feeders, often leaping from the water to catch flying insects. They are also rather undiscriminating and can be caught with a rod and line using a twist of bright cloth as bait. As their legs are esteemed as food in many places, their robust appetites are often fatal to them. Three species of Green frogs are found naturally on the European mainland and all of them have been introduced on a small scale to England.

MARSH FROG *(Rana ridibunda).* A native of southwest and eastern Europe. Its vocal sacs are grey and it is often found in large lakes and rivers. The English name comes from its most successful introduction in Britain, at Romney Marsh. Length 15cm.

POOL FROG *(Rana lessonae).* Found from France to eastern Europe and south to Italy. Its vocal sacs are usually white and, in areas where the Marsh frog lives, it prefers small pools and may spend some time on land. Length 9cm.

EDIBLE FROG *(Rana esculenta).* Very similar to the Pool frog and found in the same area; it probably arose from interbreeding between this species and the Marsh frog. Length 12cm.

AMERICAN BULLFROG *(Rana catesbeiana).* Abundant in eastern North America, this huge frog was introduced into the Po Valley of Italy over 30 years ago and now thrives there. It lives in or near water and emits a deep groaning call, a slow 'br-wum' heard especially at night. Length 15cm.

MOOR FROG *(Rana arvalis).* This species, like the others on the opposite page, is one of a group known as Brown frogs. Six of these are found in Europe and all look very alike, nearly always having a dark 'mask' on each side of the head. Most Brown frogs are much less aquatic than the Green frogs (page 78) and often only visit ponds in the spring to breed. They also differ in lacking balloon-like vocal sacs at the corners of the mouth and the calls of their males are rather weak.

The Moor frog is absent from much of western and southern Europe but elsewhere is common in very moist, open habitats such as water meadows, fens and bogs. Some animals have a stripe on the back and the proportion of these varies from place to place. Breeding males often develop sky blue patches. Length 8cm.

AGILE FROG *(Rana dalmatina).* One of our most elegant frogs, it is found in meadows and moist woods where its buff-brown colouring often matches the dead leaves among which it sits. Longer legged than its relatives, the Agile frog often makes leaps of two metres or more. Length 9cm.

COMMON FROG *(Rana temporaria).* Over much of north and central Europe this is truly the common frog and, with the Common toad, is one of the first amphibians most people come across. Where they are found with Moor frogs, Common frogs tend to occupy rather drier habitats than this species, but elsewhere they range from woods and gardens to really wet bogs and mountain pastures. They are extremely variable in appearance and even frogs in the same breeding pond may be buff, grey or reddish. Length 10cm.

On their own, *frog* and *toad* are rather vague terms and which one is used for a given species really depends only on superficial appearances. Thus *frog* is generally employed for agile, smooth-skinned animals, while *toad* is reserved for the slower, warty forms.

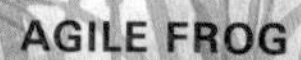
AGILE FROG
MOOR FROG
COMMON FROG

COMMON TREE FROG *(Hyla arborea).* An agile frog that climbs in trees and bushes using adhesive pads on the tips of its fingers and toes and often leaps at insect prey as they fly past. Males have a large vocal sac beneath the chin that balloons out when they produce their fast, quacking croak. In spring, tree frogs gather in ponds in huge numbers and their rolling nocturnal chorus can be heard a kilometer away. Length 5cm.

STRIPELESS TREE FROG *(Hyla meridionalis).* A southwestern form that has a much slower call than the previous species. Length 5cm.

PARSLEY FROG *(Pelodytes punctatus).* This relative of the spadefoots (p. 84) can also climb but not as well as tree frogs. It is strictly nocturnal outside the breeding season and the call may sound like a squeaky leather shoe. Length 5cm.

PAINTED FROG *(Discoglossus pictus).* Rather like a Brown frog (p. 80) but without a dark 'mask' on the face. It is active by day and night in and around shallow water. Length 7cm.

YELLOW-BELLIED TOAD *(Bombina variegata).* This common south European toad lives in shallow water from where its low musical call, 'poop . . . poop . . . poop', can be heard in spring and summer, both in the day and after dark. If another animal is unwise enough to molest it, the limbs and head are thrown up to show the bright belly, a sign that the toad is not good to eat. Length 5cm.

FIRE-BELLIED TOAD *(Bombina bombina).* Very like the previous species but the belly is often red and the finger-tips dark. It lives in the lowlands of eastern Europe. Length 5cm.

COMMON TREE FROG

STRIPELESS TREE FROG

PARSLEY FROG
PAINTED FROG
YELLOW-BELLIED TOAD
FIRE-BELLIED TOAD

MIDWIFE TOAD *(Alytes obstetricans).* A characteristic sound of summer nights in southwest Europe is the high, bell-like 'poo . . . poo . . . poo . . .' of Midwife toads which often live in piles of stones and in the crevices of walls. Unlike nearly all other European frogs and toads, these secretive animals care for their eggs which are laid in strings that the male wraps round his hind legs and carries until they are ready to hatch into tadpoles. Length 5cm.

WESTERN SPADEFOOT *(Pelobates cultripes).* There are three kinds of spadefoot in Europe, all living for preference in sandy areas where they burrow deeply and almost vertically, digging with a big spade-like tubercle on each heel and kicking the sand upwards over their head. Spadefoots emerge mainly at night, especially in wet weather. The Western spadefoot lives only in Spain, Portugal and southwestern France but further east is replaced by the Common spadefoot (*P. fuscus*) which, when molested, screams like an angry kitten and jumps at its tormentor. Length 10cm.

GREEN TOAD *(Bufo viridis).* Abundant over most of Europe except the west, the Green toad is largely nocturnal, but may sometimes be active by day, especially when young. It tolerates human contact well and often moves into outbuildings. The call is a musical liquid trill. Length 10cm.

NATTERJACK *(Bufo calamita).* This replaces the Green Toad in western areas, although the two live together in central Europe. Sandy places, including heaths and dunes are preferred and the Natterjack often runs instead of crawling or hopping. The call is much harsher than that of the Green Toad and has been likened to a football rattle. Length 8cm.

COMMON TOAD *(Bufo bufo).* More widespread than any other European amphibian, it is also the largest; in southern Europe, old females may be 15cm long and, when they inflate themselves with air, nearly as wide. The Common toad is found in all sorts of fairly dry places including town gardens. It is quite intelligent and soon becomes tame. One is known to have lived at least 36 years. The call is a weak 'kwark kwark'. Length 15cm.

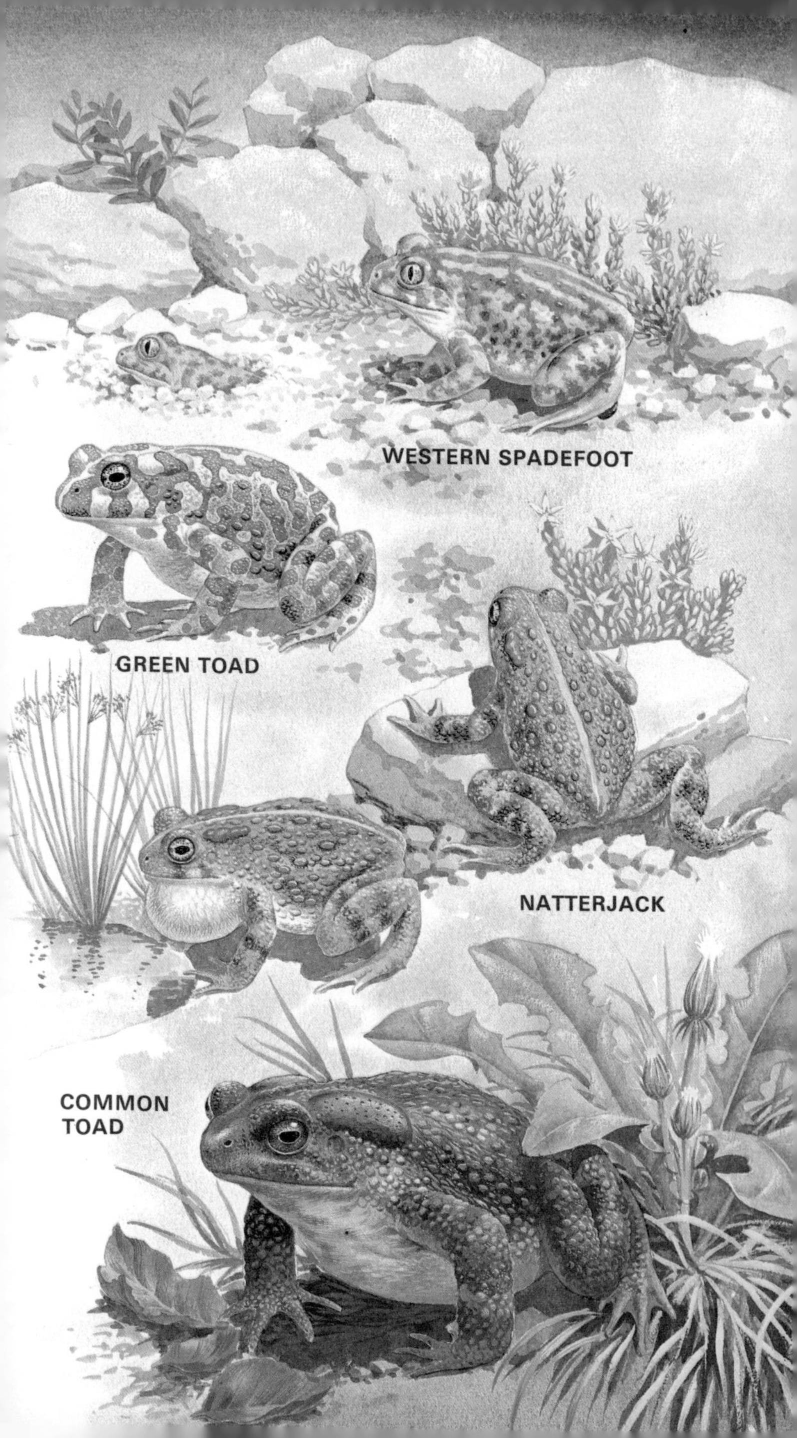
WESTERN SPADEFOOT
GREEN TOAD
NATTERJACK
COMMON
TOAD

HERMANN'S TORTOISE

MARGINATED TORTOISE

HERMANN'S TORTOISE *(Testudo hermanni).* Although mainly vegetarian, tortoises will also feed on carrion and even the dung of other animals. They live in a variety of places with good plant cover, ranging from dry hillsides and sand dunes to lush meadows and cultivated land and even rubbish dumps. Most activity takes place in the mornings and late afternoons. Females lay up to a dozen round, white, hard-shelled eggs in dry soil. Hermann's tortoise lives mainly in Italy and southeast Europe. The very similar Spur-thighed tortoise *(Testudo graeca)* is found in the eastern Balkans and Spain and has often been imported into Britain from Morocco and Turkey to be sold in pet shops. Length 20cm.

MARGINATED TORTOISE *(Testudo marginata).* Found only in Greece and Sardinia, this animal often lives higher in the mountains than Hermann's tortoise. The population on Sardinia is said to result from a custom of Roman times: seamen allegedly brought Marginated tortoises from Greece as presents for their intended wives. Length 30cm.

EUROPEAN POND TERRAPIN *(Emys orbicularis).* Terrapins can be distinguished from tortoises by their flatter, smoother shell and the webbing between their toes. They are almost always found in or near water where they swim and dive with great skill. Various small animals including fish, frogs and even ducklings are eaten. The European pond terrapin is a shy and retiring animal that prefers still or slow-flowing water with a good growth of aquatic plants or overhanging vegetation. Length 20cm.

STRIPE-NECKED TERRAPIN *(Mauremys caspica).* A more southerly species than the European pond terrapin, this animal is often found in more open waters and tolerates brackish conditions, occasionally being seen sunning itself on mud banks quite near the sea. Length 20cm.

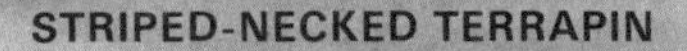

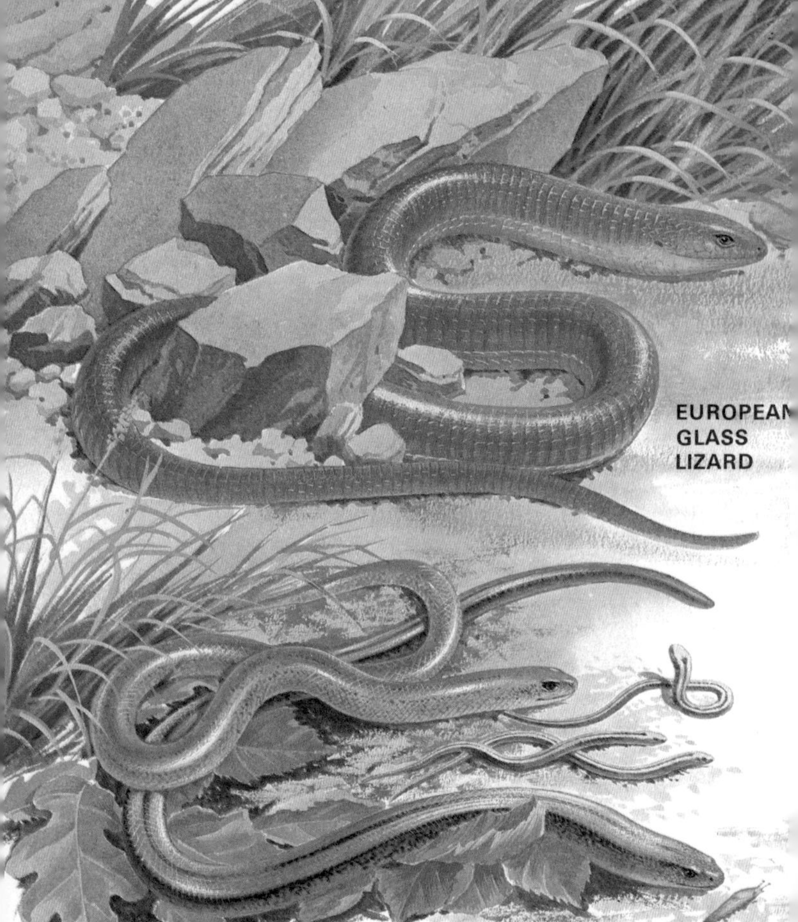

GLASS LIZARD *(Ophisaurus apodus).* This is the largest and heaviest of the European lizards, growing as long as 120cm and as thick as a man's wrist. It is now found only in southeastern Europe and southwest Asia, but fossils show that Glass lizards once reached as far north as Germany. At first sight, the Glass lizard is very snake-like, but this and some other more or less limbless lizards differ from snakes in several ways, for instance they have eyelids that can be closed and tails that can be broken off but will grow again. The English name was first applied to American Glass lizards and refers to their shiny appearance and the way their fragile tails break into pieces if struck; in the European form breakage does not take place so easily. These animals are found in quite dry areas, especially near stone piles and old walls, where they feed on large invertebrates and even other lizards and small mammals. Length 120cm.

SLOW WORM *(Anguis fragilis).* A small relative of the Glass lizard that lives over most of Europe, especially in rather moist places where it is often discovered under flat stones. Large numbers of slugs are eaten and females produce about six to twelve fully formed young in late summer. A slow worm has survived 54 years in captivity. Length 50cm.

THREE-TOED SKINK *(Chalcides chalcides).* Extremely fast, in spite of its tiny limbs, this skink can scarcely be seen as it rushes through herbage in meadows in Italy and southwest Europe. Up to 23 fully formed young are produced. Length 40cm.

OCELLATED SKINK *(Chalcides ocellatus).* A shy ground lizard found in a few often sandy areas of southern Europe. Length 30cm.

AMPHISBAENIAN *(Blanus cinereus).* Found in Spain and Portugal where it burrows in sandy soil and rarely comes to the surface. This reptile is neither a snake or a lizard but is a relative of both. Length 30cm.

CELLATED SKINK

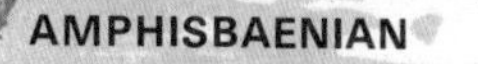

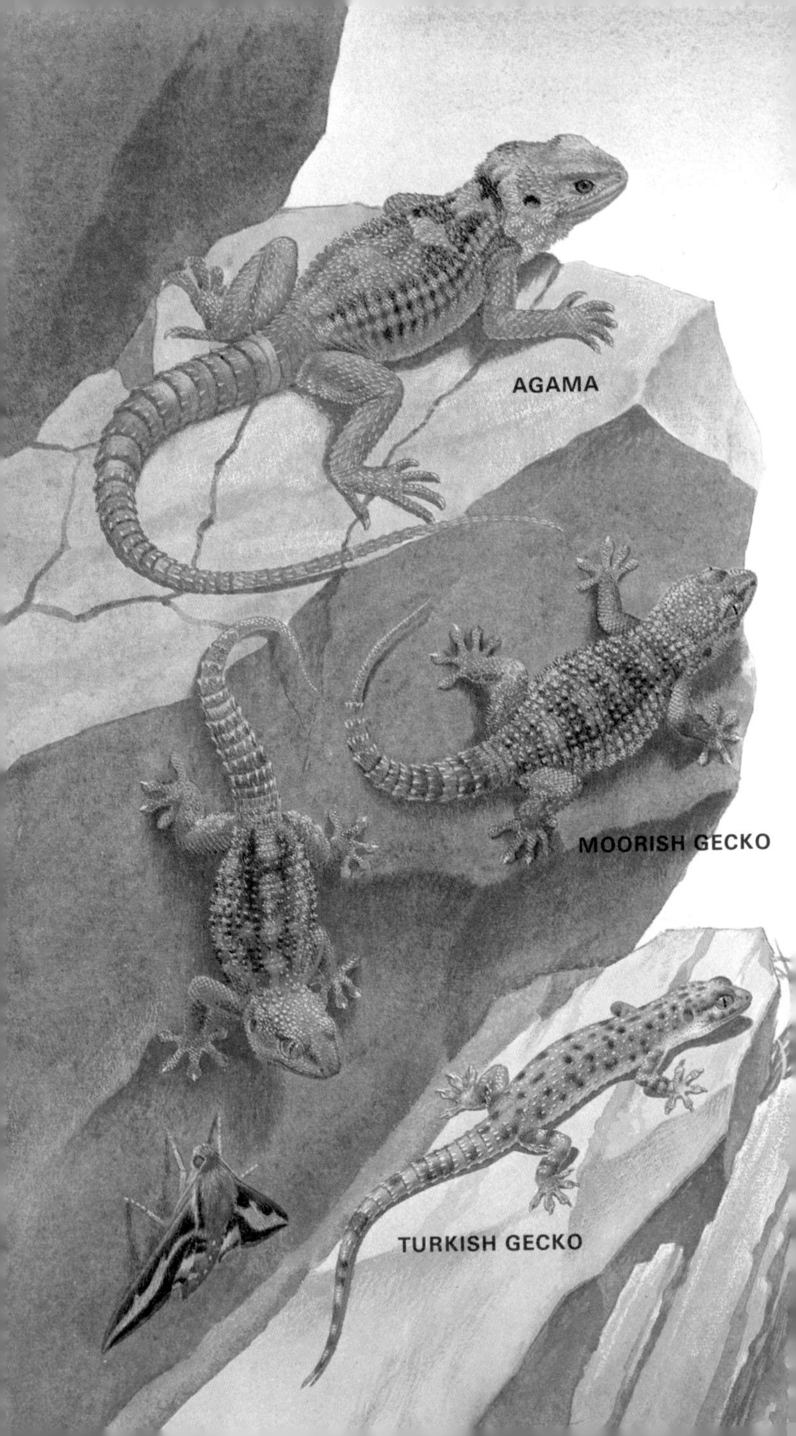
AGAMA
MOORISH GECKO
TURKISH GECKO

CHAMELEON

AGAMA *(Agama stellio).* An inhabitant of parts of Greece and some Aegean islands, the Agama lives on walls, trees and the roofs of low buildings and characteristically nods its head at intervals. It is an efficient climber that is active by day and is capable of some colour change, although not as much as the chameleon. Length 30cm.

MOORISH GECKO *(Tarentola mauretanica).* Geckoes live in the warmer parts of the world including the Mediterranean area. Most are nocturnal but the Moorish gecko is quite often seen by day, especially in the cooler parts of the year. Like many of its relatives, it has large eyes with vertical pupils, and flattened adhesive pads on the fingers and toes that allow it to climb with great agility, even on the smooth walls and ceilings of houses, which it often enters. Length 15cm.

TURKISH GECKO *(Hemidactylus turcicus).* More nocturnal than the Moorish gecko, this species is often rather translucent. As with most other geckoes it lays eggs that are hard-shelled like those of birds, not leathery as in other lizards. Length 10cm.

CHAMELEON *(Chamaeleo chamaeleon).* A member of a largely African and Madagascan family that reaches our area only in southern Spain and Crete, it is beautifully adapted for living in trees and bushes and has a prehensile tail and toes that are arranged for grasping twigs and branches. Good camouflage is produced by the flat, leaf-shaped body and ability to change colour, although chameleons do not always match their background. These animals are very slow-moving and stalk their insect food painstakingly before catching it on their long, sticky tongue which can be shot forwards for almost the length of the lizard. Length 30cm.

Δ

DALMATIAN ALGYROIDES *(Algyroides nigropunctatus)*. Often seen on field walls, where it can be quite conspicuous against the pale stones it also spends a great deal of time climbing in bushes among which it is very difficult to locate. On Corfu, where there are no climbing Wall lizards, the Dalmatian algyroides becomes the common lizard of towns and villages. Close relatives live in Greece, Corsica and Sardinia and southeast Spain. Length 20cm.

LARGE PSAMMODROMUS *(Psammodromus algirus)*. A common lizard in much of Spain and Portugal and parts of France that is particularly abundant in sandy places with dense, prickly bushes and shrubs. It spends much of its time hunting around the bases of these, but also climbs in them, the large, pointed scales that cover its body protecting the lizard against their spines. Like geckoes, it has a voice and often squeaks when picked up. Length 25cm.

SPANISH PSAMMODROMUS *(Psammodromus hispanicus)*. Often found among low, mat-forming plants, where it may be glimpsed dodging from clump to clump, although near the coast it is sometimes encountered in open sandy or gravelly places. The Spanish psammodromus is one of the few European lizards that regularly breeds the year after it hatches, most others taking at least two seasons to reach maturity. Length 13cm.

SPINY-FOOTED LIZARD *(Acanthodactylus erythrurus)*. An inhabitant of sandy areas and other open places in Spain and Portugal where it may run long distances with the tail raised in a gentle curve. Frightened animals often take refuge in burrows that they dig themselves. The strikingly coloured young frequently wave their bright red tails. Length 22cm.

LARGE PSAMMODROMUS
SPANISH PSAMMODROMUS
SPINY-FOOTED LIZARD
Juvenile

OCELLATED LIZARD
Juvenile
Juvenile
♂
GREEN LIZARD
♀

OCELLATED LIZARD *(Lacerta lepida).* One of the largest European lizards, this fine animal is abundant in Spain and Portugal and parts of southern France. Shy but adaptable, it survives in a wide range of habitats including arable land, vineyards, woods and sand dunes. Many large insects are eaten, but smaller lizards and small mammals, up to the size of baby rabbits, are also captured. In spring the Ocellated lizard raids bird nests, climbing bushes and even quite tall trees, and later in the season a substantial amount of fruit may be eaten. Females produce up to two dozen eggs. Captive Ocellated lizards have lived twenty years. Length 60cm.

GREEN LIZARD *(Lacerta viridis).* A relative of the previous species, it lives in bushy places such as bramble patches where it may be seen basking at the edges of the clumps. Green lizards climb well, even in flimsy vegetation, using their long tails to spread their weight. Length 40cm.

SAND LIZARD *(Lacerta agilis).* A familiar lizard over much of Europe that spends its time on the ground or in low vegetation. Colouring is very variable and there are regional differences, for instance, in parts of eastern Europe, some males are virtually all green and in Britain animals with red-brown backs are not found; breeding males have brilliant grass-green sides. In northwestern Europe, this species is becoming rare and is found only in sandy areas, hence its name. It is believed that, in regions with cool summers, open sun-warmed sand provides the necessary heat to hatch the buried eggs. Further south, Sand lizards are more widespread and may be found in dry meadows and even kitchen gardens. Length 22cm.

♀

SAND LIZARD

♂

♀

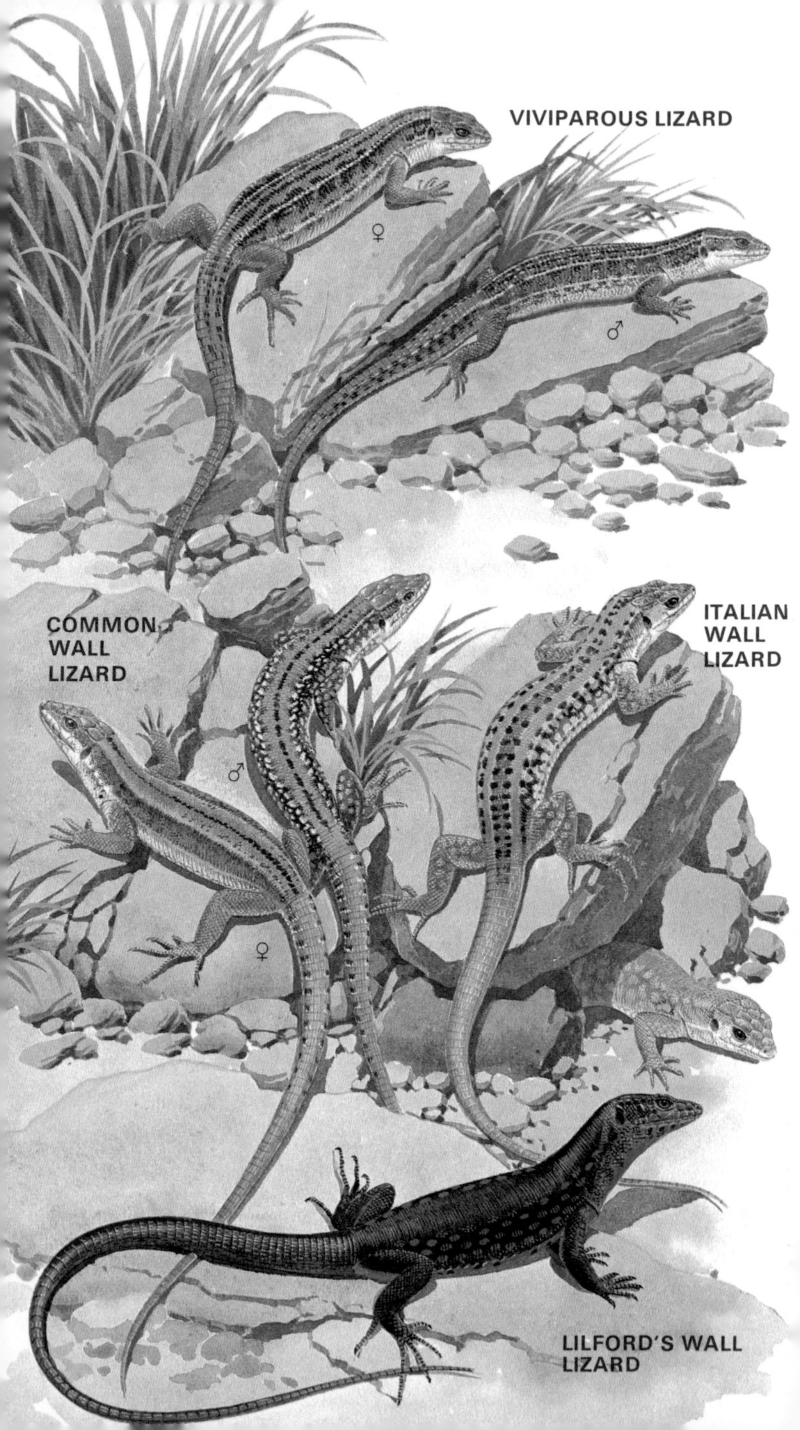
VIVIPAROUS LIZARD
♀
♂
COMMON
WALL
LIZARD
♂
♀
ITALIAN
WALL
LIZARD
LILFORD'S WALL
LIZARD

VIVIPAROUS LIZARD *(Lacerta vivipara).* Abundant in cool, moist places over much of Europe, this little lizard is found well within the Arctic circle and eastwards to the Pacific coast of the USSR and is the only reptile to reach Ireland. It lives on heaths, marshes and railway embankments and in open woods, hedge bottoms and gardens, but in the south is confined to mountains and very wet places such as rice fields. Females do not usually lay eggs, blackish-bronze babies being produced in late summer in broods of from two to eleven. Length 17cm.

COMMON WALL LIZARD *(Podarcis muralis).* Wall lizards, of which there are 14 species, are found only in Europe and nearby areas. Most of them live near the Mediterranean, but the Common wall lizard is more widespread. It likes drier places than the Viviparous lizard and is a much better climber being seen on walls, rocky banks, trees and telegraph poles and even in the uppermost twigs of hedges. An intelligent, adventurous lizard, it is quick to explore new habitats and to seize opportunities. Like geckoes, Wall lizards invade human settlements, although they do not usually enter houses. Eggs are produced in broods of two to four, sometimes three times a year. Length 20cm.

ITALIAN WALL LIZARD *(Podarcis sicula).* In spite of its name, this wall lizard occurs in some other areas. Often bigger, and always more aggressive than the Common wall lizard, it is not quite such a good climber but can run long distances across open ground. Length 25cm.

LILFORD'S WALL LIZARD *(Podarcis lilfordi).* Confined to barren islets off Majorca and Minorca, some populations are brown or green rather than black. The sparse insect diet available is eked out with plant material. Length 22cm.

SHARP-SNOUTED ROCK LIZARD *(Lacerta oxycephala).* The most agile of the six rock lizards in Europe, it climbs high on cliffs, walls and fortifications in southeast Yugoslavia. Here it is liable to attack by birds and the bright tail attracts their attention to a part of the lizard that is dispensable. Length 18cm.

▷

△

ADDER *(Vipera berus).* This is one of the five kinds of viper that are at all widespread in our area. Vipers are the only European snakes that are really dangerous to man and must all be treated with respect. Like the others, the Adder produces fully developed young and feeds largely on small mammals and lizards, the prey usually being killed by the long poison fangs which are placed at the front of the upper jaw and fold back out of the way when not in use. It prefers rather open habitats and ranges further north than any other snake, being found even within the Arctic Circle and across the USSR to the Pacific Ocean. The colouring of Adders varies considerably and some are even black, but males generally tend to be grey, and females brownish or reddish. In the spring, the males may take part in combat 'dances,' two of them rearing up and pressing against each other until the weaker one retires. The females, in at least some areas breed only once every two years. Length 65cm.

ORSINI'S VIPER *(Vipera ursinii).* Strangely, this little viper is quite docile and rarely bites people even when molested. The head is narrower than that of the very similar Adder and the skin looks rougher. It is only found in a few small areas of southern Europe, favouring meadowland and hillsides with low vegetation. Some populations eat a lot of grass-hoppers. Length 50cm.

ASP VIPER *(Vipera aspis).* Replacing the Adder over much of Italy and France, the Asp viper varies greatly in pattern but can be identified by its turned up snout. Length 60cm.

NOSE-HORNED VIPER *(Vipera ammodytes).* This is the common viper of much of southeastern Europe showing a preference for dry rocky areas where it is sometimes seen working its way along cliff ledges looking for lizards. Like other vipers, it is largely diurnal but may also be active by night where the climate is mild enough. The 'horn' on its snout is quite soft and flexible. The very similar Lataste's viper *(Vipera latasti)* is found in Spain and Portugal. Length 65cm.

ADDER

ASP VIPER

NOSE-HORNED VIPER

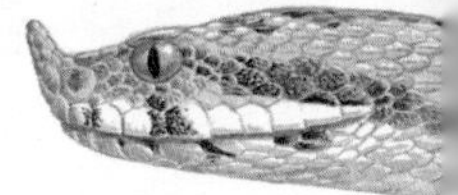

ORSINI'S VIPER

ASP VIPER

NOSE-HORNED VIPER

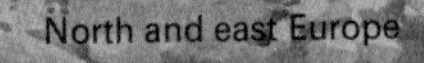

△

GRASS SNAKE *(Natrix natrix).* This snake is found over most of Europe except the far north. Usually diurnal, it hunts frogs and toads but also eats newts and fish at times. It is often seen near water where it swims well but in cooler countries may be encountered in quite dry places as well. Grass snakes rarely actually bite but often bluff if captured, hissing and shooting their heads forwards as if they were going to do so; they may also void the smelly contents of their anal glands or even pretend to be dead. Eggs are laid in rotting vegetation, the warmth of the decomposing plants speeding development. Length 120cm.

VIPERINE SNAKE *(Natrix maura).* Common in southwest Europe, this relative of the Grass Snake has similar habits but spends more time in water and eats a higher proportion of fish. Some animals have a dark zig-zag stripe along the back and can look very viper-like, especially when they are partly hidden by plants. Length 100cm.

DICE SNAKE *(Natrix tessellata).* The Viperine snake is replaced by this species in much of Italy and southeast and central Europe; isolated colonies also exist further north, in Germany. The Dice snake is even more aquatic than the previous species and takes even more fish. It can remain submerged for long periods and it is often possible to watch an individual crawling along the bottom of a stream in a leisurely way, pushing its nose under every large pebble as it searches for small fish that might lurk there. Length 100cm.

Southeast Europe
GRASS
SNAKE
VIPERINE SNAKE
DICE SNAKE

WESTERN WHIP SNAKE *(Coluber viridiflavus).* Whip snakes are among the fastest serpents in Europe and are active by day, hunting largely by sight. Although not venomous they bite vigorously if captured. The Western whip snake is common in Italy and much of France in a variety of rather dry places that usually have plenty of vegetation. Lizards form a large part of the diet but mammals, birds and even other snakes are also eaten. The smaller but closely related Balkan whip snake *(Coluber gemonensis)* takes a large number of grasshoppers. Babies develop adult colouring when they are about three years old and in Italy many animals become almost completely black. Length 150cm.

HORSESHOE WHIP SNAKE *(Coluber hippocrepis).* Found in Spain, Portugal and Sardinia, this whip snake has similar habits to the preceding species but is more frequently seen close to human settlements. The name refers to a dark mark on the back of the head that is sometimes shaped like a horseshoe. Length 150cm.

LARGE WHIP SNAKE *(Coluber jugularis).* This is one of the largest European snakes and is common in the eastern Balkans where it feeds mainly on small mammals. Unlike most snakes, it is sometimes bold enough to stand its ground when approached. Length 200cm.

DAHL'S WHIP SNAKE *(Coluber najadum).* The smallest and quickest of our whip snakes, it prefers dry, rocky places with bushes and herbage in which it climbs with great verve and dexterity. Food consists of small lizards and grasshoppers. Length 135cm.

MONTPELLIER SNAKE *(Malpolon monspessulanus).* Common in much of southern Europe except Italy, this snake specializes in hunting lizards by sight. Prey, which may include even full-grown Ocellated lizards, is killed by venom injected by poison fangs at the back of the upper jaw. Length 200cm.

MONTPELLIER SNAKE

LEOPARD SNAKE *(Elaphe situla).* Like the others shown here, this species is one of a group known as rat snakes and, as the name suggests, feeds on small mammals when adult, although the babies eat many lizards. Leopard snakes live in stony places in southeast Europe and southern Italy, and may sometimes be encountered in the crevices of field walls and around barns and houses where they are attracted by the rodents living there. Length 100cm.

LADDER SNAKE *(Elaphe scalaris).* So named because the young often have a dark, ladder-shaped pattern on the back which is replaced by two simple lines in adults. A fast, day-active snake, it climbs well on walls and bushes in dry, sunny places such as vineyards and warm hillsides in Spain, Portugal and southern France. A mature animal can capture prey up to the size of a half-grown rabbit; as with other rat snakes, large prey is killed by constriction (see p. 107). Length 160cm.

LEOPARD SNAKE

LADDER SNAKE

Juvenile

AESCULAPIAN SNAKE *(Elaphe longissima).* The most widespread European rat snake and the best climber, sometimes confidently ascending vertical tree trunks to raid bird nests although small mammals are the largest item of diet. It is often found along wood edges but turns up in all sorts of situations including even the tops of hay stacks. Although mainly a southern species, there are a few isolated colonies in Germany and Czechoslovakia in warm, sheltered places. Length 150cm.

FOUR-LINED SNAKE *(Elaphe quatuorlineata).* In spite of the name, this snake is not always striped, for babies have a blotched pattern which is kept by adults in the eastern Balkans, but in the rest of southeast Europe and Italy the blotches give way to lines when the snakes are about three years old. One of the largest European reptiles, it is rather slow-moving and likes a humid atmosphere, often being found near water and hunting in cloudy weather and at dusk. The Four-lined snake is found on many Aegean islands but rarely on those inhabited by the Large whip snake. Length 250cm.

CAT SNAKE
SMOOTH SNAKE
SAND BOA
WORM SNAKE

CAT SNAKE *(Telescopus fallax).* Common in stony areas in the warmer parts of the Balkans, this snake is mainly active in the evening when it hunts small lizards. These are pulled from their hiding places or stalked in the open, the snake creeping from cover to cover before making a final rush; a habit which, together with its slit-shaped pupils in good light, is the source of this reptile's name. Prey is quickly killed by venom injected by fangs at the back of the upper jaw. Length 75cm.

SMOOTH SNAKE *(Coronella austriaca).* One of the more widespread European snakes, it reaches southern England but is only found there in a few areas of dry heathland. Further south, Smooth snakes also live in open woods, hedgebanks and rocky places. Like Cat snakes they eat a lot of lizards, holding them in the coils of their body to prevent their escape. Females give birth to two to fifteen young at a time. Length 65cm.

SAND BOA *(Eryx jaculus).* Most snakes have no sign of any limbs, but the Sand boa has a 'claw' on each side of the vent that represents the hind leg. Such vestiges are found in many of its relations, such as Pythons, Anacondas and Boa constrictors. Found in southeast Europe, this snake is easily recognised by its plump, glossy body and rounded tail tip. It may spend the day buried in sandy soil and emerges in the evening to hunt small lizards and mammals, often pursuing the latter into their burrows. Like their bigger relations, Sand boas constrict prey, tightening the coils of the body until the victim's breathing is stopped. Fully developed young are produced. Length 80cm.

WORM SNAKE *(Typhlops vermicularis).* This, the smallest European snake, looks more like a dry, shiny worm. Its tiny eyes are placed beneath the head scales and the very short tail ends in a spine. Worm snakes lead a very secretive existence and are rarely seen above ground, spending most of their time in burrows but they can sometimes be found by turning over stones in meadows and on grassy slopes in southeast Europe. Their diet consists mainly of small insects. Length 30cm.

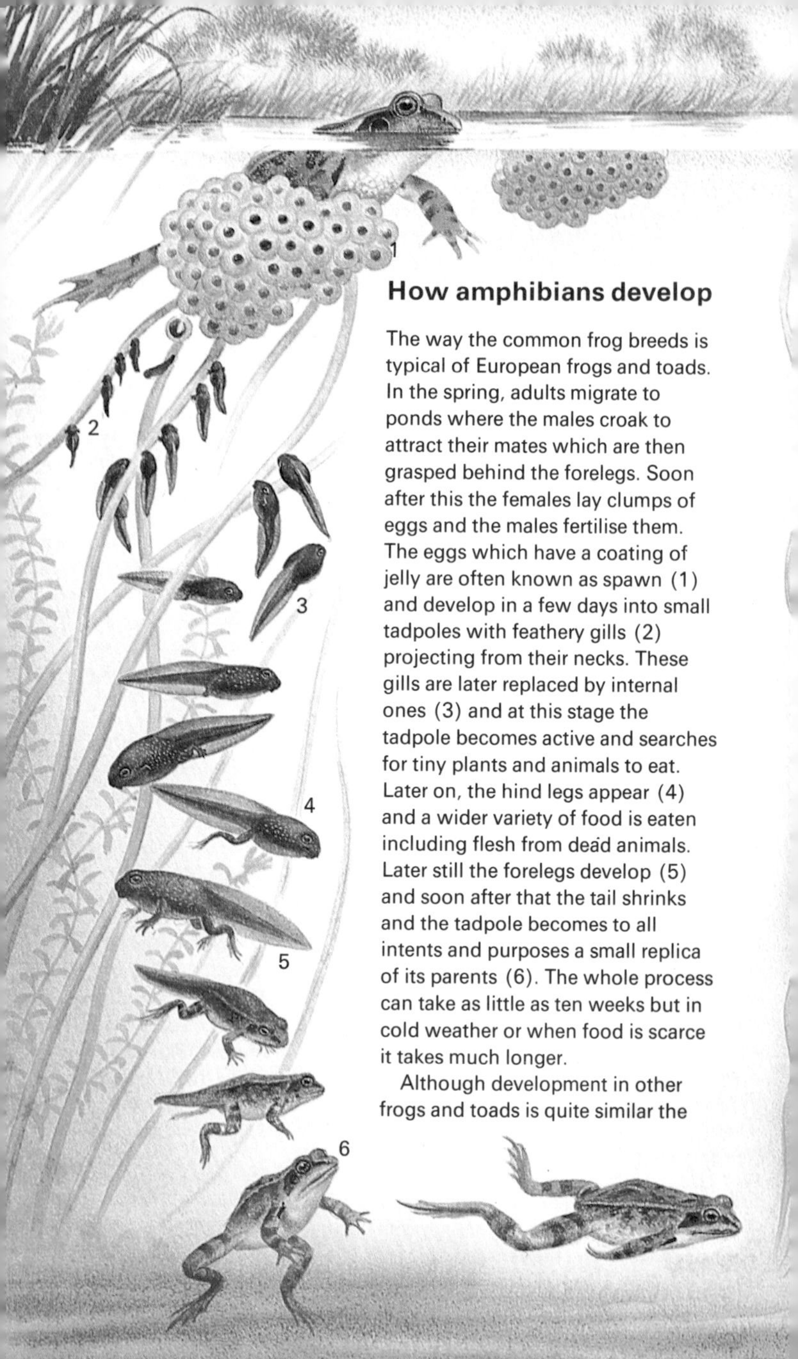

How amphibians develop

The way the common frog breeds is typical of European frogs and toads. In the spring, adults migrate to ponds where the males croak to attract their mates which are then grasped behind the forelegs. Soon after this the females lay clumps of eggs and the males fertilise them. The eggs which have a coating of jelly are often known as spawn (1) and develop in a few days into small tadpoles with feathery gills (2) projecting from their necks. These gills are later replaced by internal ones (3) and at this stage the tadpole becomes active and searches for tiny plants and animals to eat. Later on, the hind legs appear (4) and a wider variety of food is eaten including flesh from dead animals. Later still the forelegs develop (5) and soon after that the tail shrinks and the tadpole becomes to all intents and purposes a small replica of its parents (6). The whole process can take as little as ten weeks but in cold weather or when food is scarce it takes much longer.

Although development in other frogs and toads is quite similar the

1 2 3 4 5 6

Smooth newt

shape of the egg masses is very variable, for instance Common, Green and Natterjack toads produce long strings instead of clumps. The tadpoles also differ considerably in size: in the case of the large Common toad they are only 3.5cm long while the smaller Spadefoots' tadpoles can grow to 16cm with bodies the size of hens' eggs. There are also differences in diet and in agility, most tadpoles being relatively slow while those of tree frogs can dart swiftly about like fish.

The main stages in the development of a Smooth newt are shown at the top of this page. Here too the eggs are laid in water but they are produced singly and carefully wrapped in the leaves of plants. The tadpoles, unlike those of frogs and toads, keep their feathery gills until they leave the water and the forelegs develop before the hind ones. Brook, Sharp-ribbed and Spectacled salamanders develop in a generally similar way but Fire salamander eggs, reach the tadpole stage within the mother's body and are then either released into water or kept safe within their parent until they are fully developed, which is what normally happens in Alpine salamanders. Cave salamanders lay their eggs away from water in damp places where they hatch fully formed.

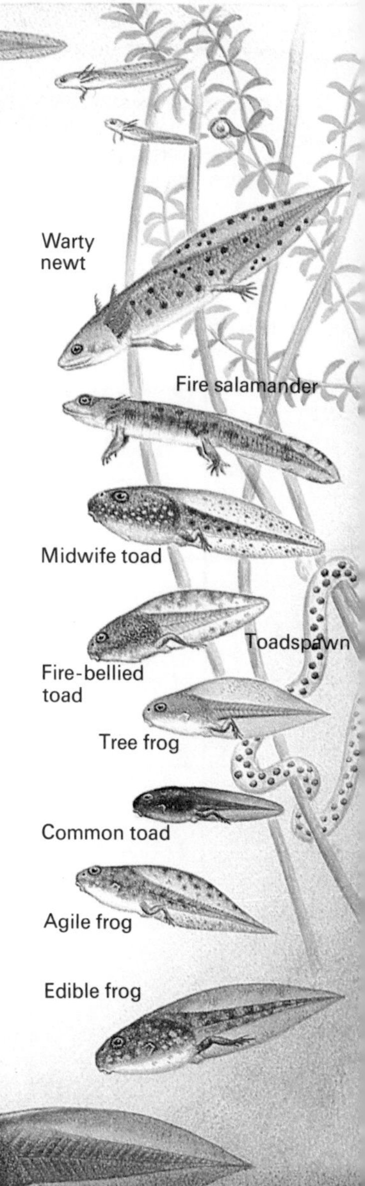

Spadefoot toad

Tracks and Traces

Watching wild animals requires skill and care but will richly reward your efforts. One of the most satisfying aspects of studying animals in the wild is the detective work necessary for interpreting tracks, droppings and feeding traces. These will help you to learn the local behaviour of an animal or group of animals, and from this you may be able to find the best places from which to keep watch.

Tracks vary a great deal from species to species, and also in the individual animal. For example, a deer walking will leave a track showing the hind foot placed in the print of the fore foot. The faster the animal moves, the further in front of the fore foot goes the hind foot. If the animal leaps, then the small hind toes register as well as the hooves, and the hooves of the fore feet are splayed outwards.

Deer tracks

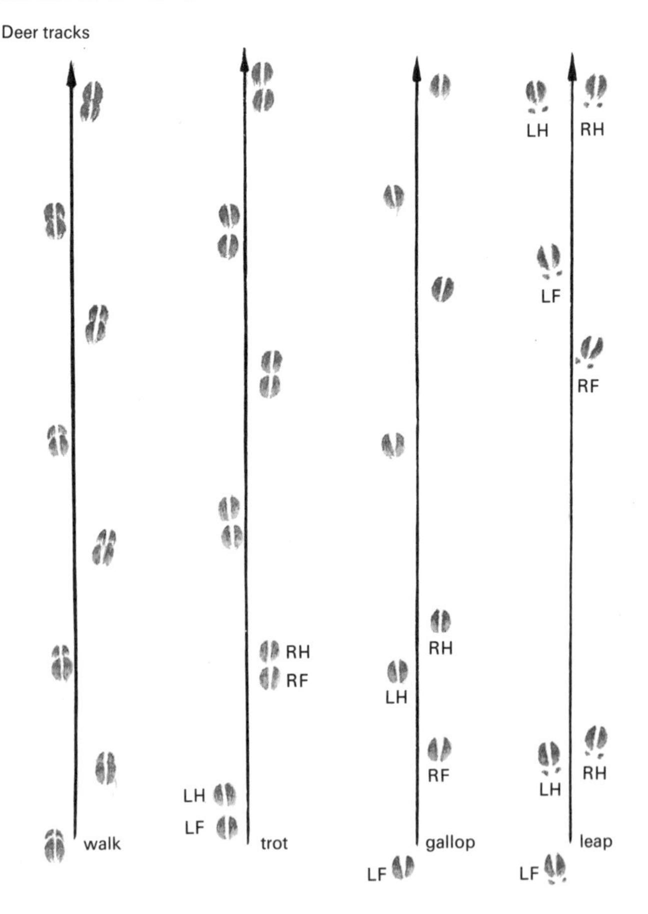

Foxes frequently walk slightly sideways, which produces a track with the fore feet on one side of an imaginary 'line of advance', and the hind feet on the other. Just to make things more difficult for the observer, they will change step and turn across this imaginary line, placing the feet on the opposite side of the line to the previous series of prints.

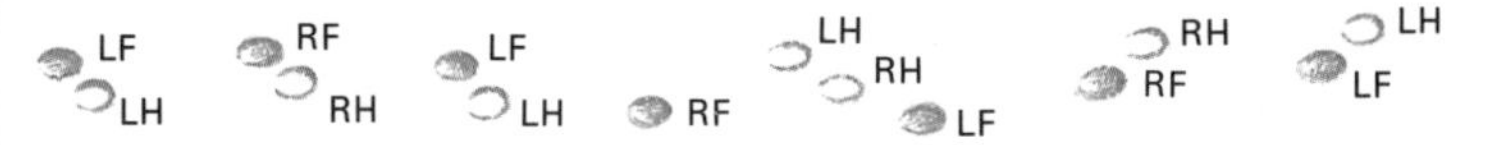

The Smallest Animals

Shrews, mice, voles and lizards leave few tracks, for they are light, and keep away from muddy surfaces, but do leave characteristic trails in snow. Field voles make easily recognized runs in grassland, with little piles of droppings, often covered with short bitten grass.

Trail of shrew from hole in snow

Jumping trail of woodmouse in snow

Running trail of bank vole, showing drag of tail

Track of field vole, showing partly covered droppings

Rodents, rabbits and insectivores

In the wild, animals will take up a particular range, or territory, the size of which is decided by the food requirements of the group or individual concerned. Such territories are nearly always marked by deposits of droppings at strategic points. One of the best and most easily observed examples of this behaviour is a rabbit colony. Here you can clearly see the patches of droppings, of all ages, since the characteristic scent of the colony must be maintained in order to warn off intruding rabbits from other colonies. At the same time you can usually spot the tracks from the burrows radiating out to each patch, sometimes on a hummock or a tree stump.

Rabbit and hare droppings are not unlike those of deer. This is not surprising, for they eat similar food. In both groups the weather eventually dries out the droppings and they break up.

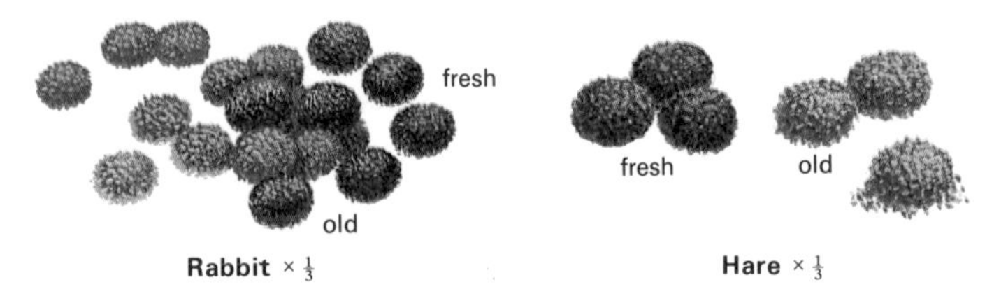

Rabbit × $\frac{1}{3}$ **Hare** × $\frac{1}{3}$

The larger rodents often leave reasonably clear tracks and, since rats, water voles, muskrats and coypu spend much time near water, muddy banks are ideal sites on which to search. The rounded cylindrical droppings of voles are very different from those of rats and mice.

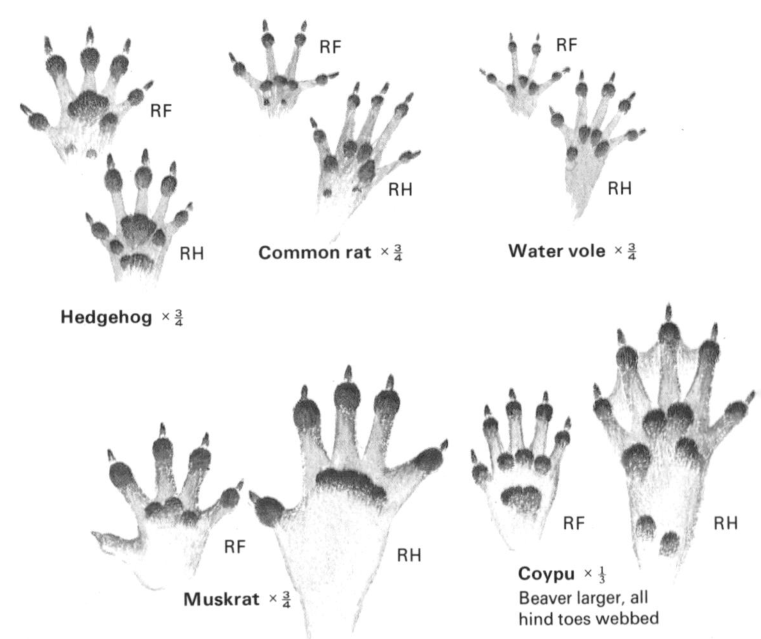

Hedgehog × $\frac{3}{4}$ **Common rat** × $\frac{3}{4}$ **Water vole** × $\frac{3}{4}$

Muskrat × $\frac{3}{4}$

Coypu × $\frac{1}{3}$
Beaver larger, all hind toes webbed

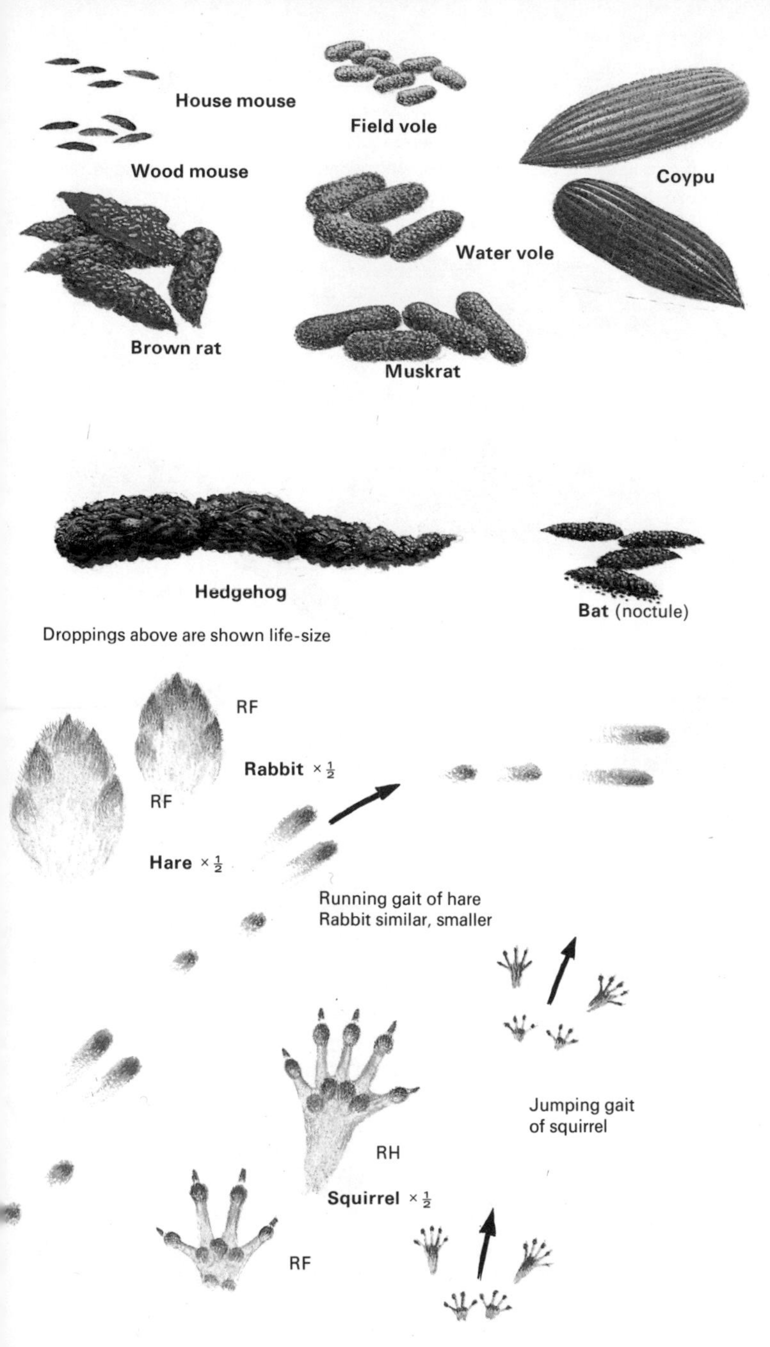

Droppings above are shown life-size

Carnivores

Droppings of carnivores are very different from those of herbivores and, when fresh, the colour bears some relationship to the food eaten. This is particularly the case with foxes, whose great variety of diet produces much variation in the colour of the droppings. Droppings containing a quantity of bony material will weather to a brittle, whitish grey.

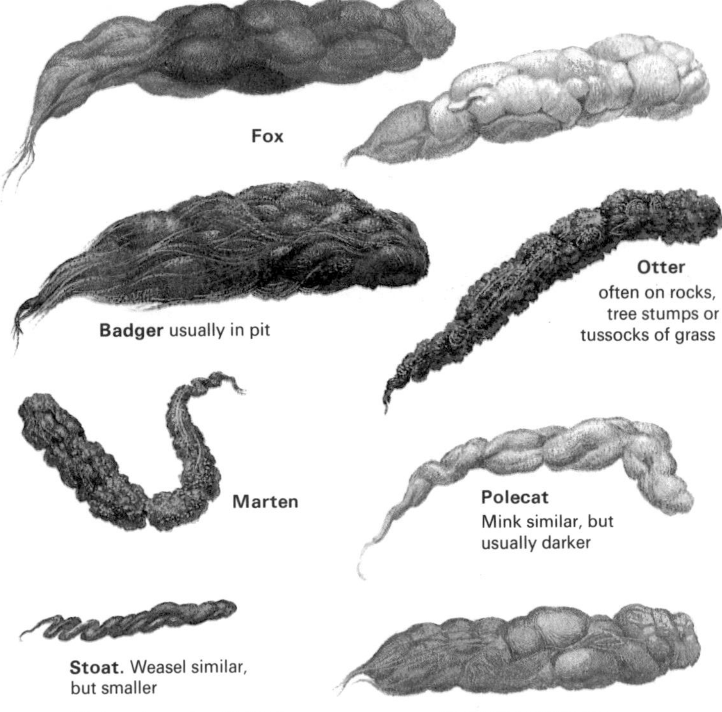

All droppings about half-size

Carnivore tracks are very distinctive in the larger species. The claws show clearly in fox tracks, together with creasing between the pad impressions, caused by the hairy nature of the feet. This is not seen in the broader impressions made by dogs.

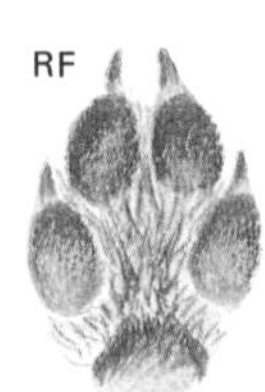

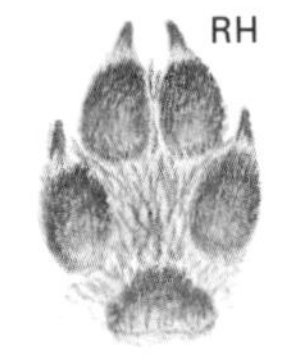

Fox $\times \frac{1}{2}$

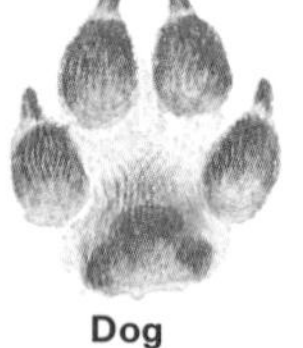

Dog

Wolf $\times \frac{1}{3}$

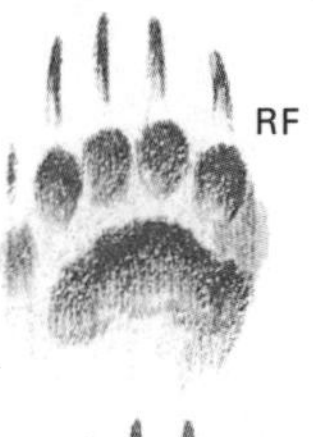

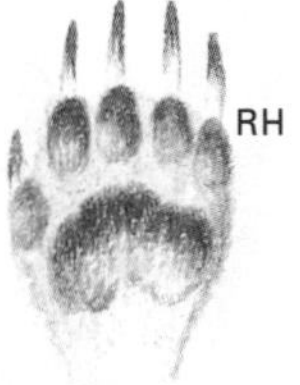

Badger × ½
Bear similar, but very large indeed

Otter × ½
Wolverine similar, but not webbed. Sole pads are smaller

Raccoon × ½

Cat claws are retractible, and do not show in the footprints.

Cat × ½

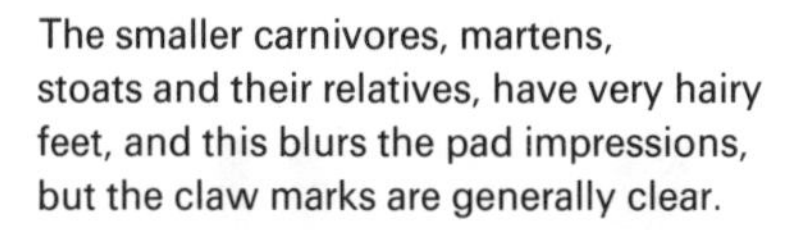

The smaller carnivores, martens, stoats and their relatives, have very hairy feet, and this blurs the pad impressions, but the claw marks are generally clear.

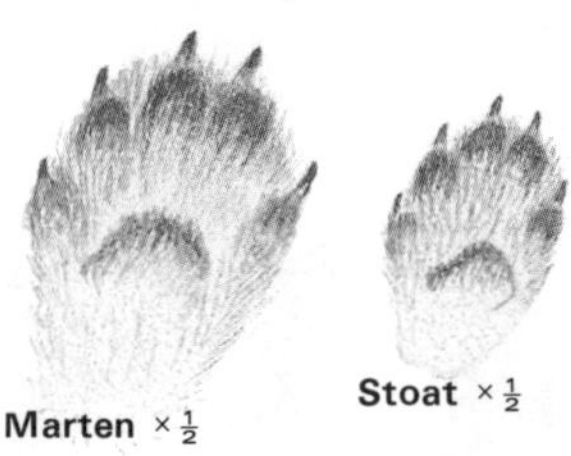

Marten × ½

Stoat × ½

Hoofed animals

Tracks of hoofed animals are generally more easily seen than those of any other group, because more weight per square centimetre rests on the foot of a deer than on that of a fox. This means that prints will appear on ground where other animals leave hardly a trace.

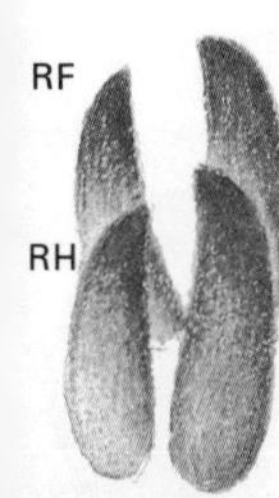

Muntjac
× ¾

inner hoof often smaller than outer

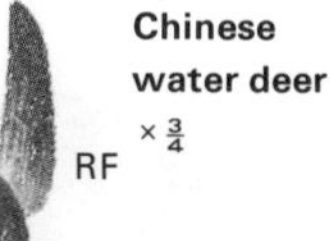

Chinese water deer
× ¾

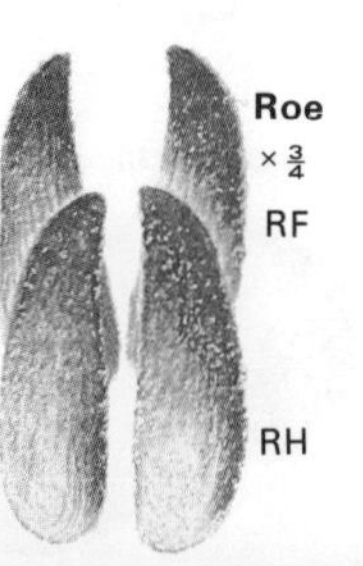

Roe
× ¾

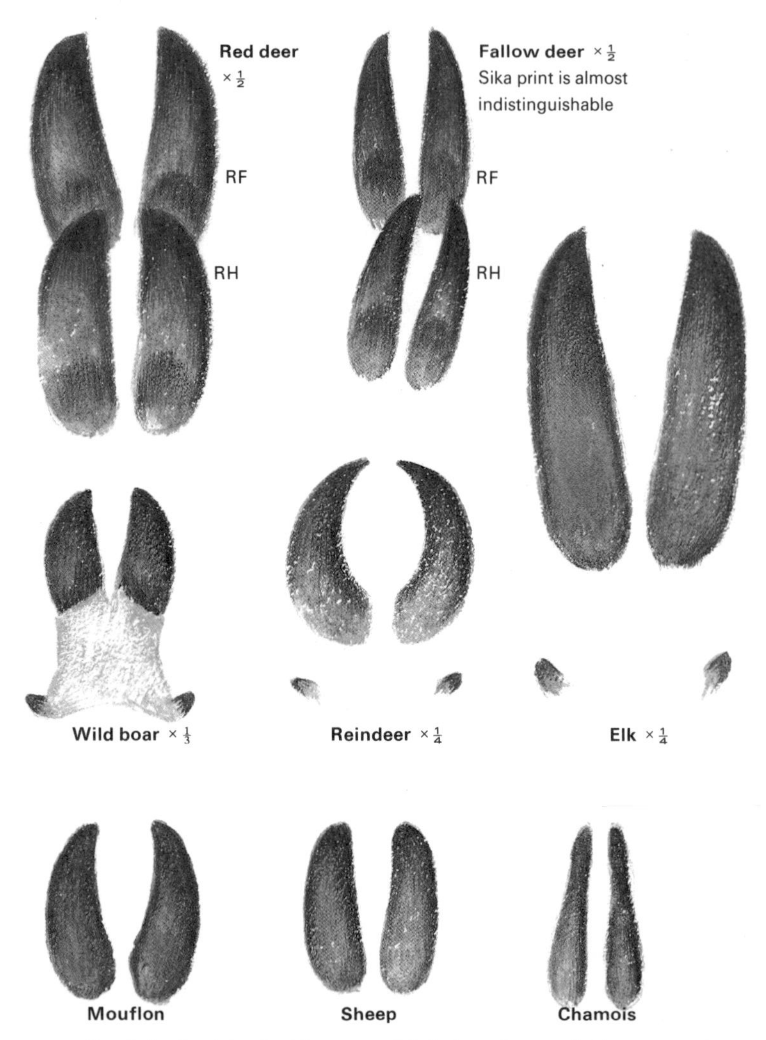

Hoofed animals are herbivores, and their droppings are fairly constant in shape, though there is sometimes variation between male and female animals. The season affects the kind of food eaten, and this dictates whether the droppings stick together in a mass or lie more or less separate.

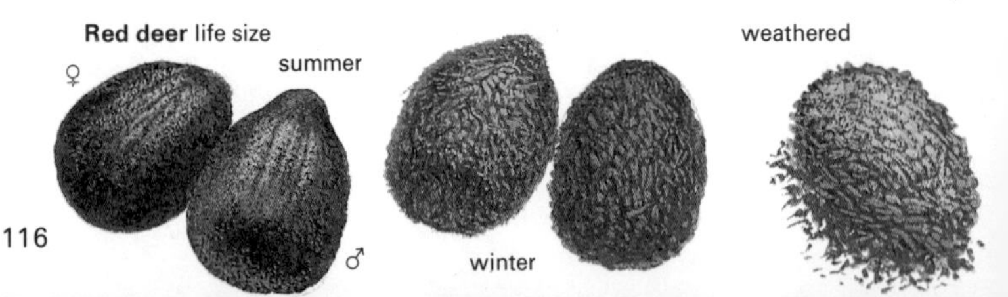

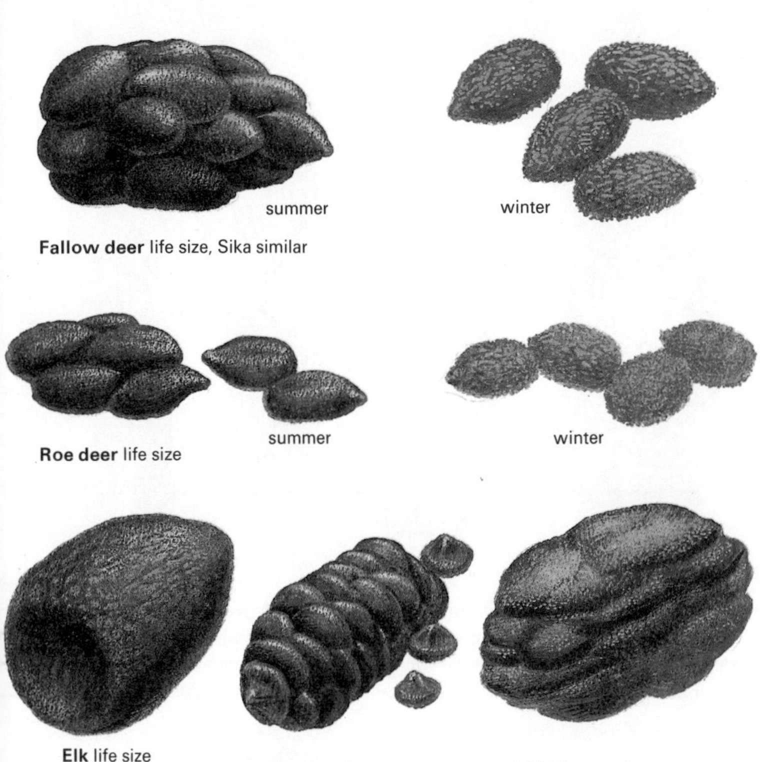

Fallow deer life size, Sika similar

Roe deer life size

Elk life size

Sheep $\times \frac{1}{2}$
mouflon similar

Wild boar $\times \frac{1}{2}$

Feeding signs

Tracks and droppings are not the only signs of animal activity. Many animals leave distinctive evidence of their feeding habits, and one of the most common is bark gnawing. In winter it is done when feeding, but summer bark gnawing is often done by voles and squirrels for no obvious reason, since the bark is stripped off and left lying about, uneaten.

Growing points and side shoots of young trees are bitten off by hares and deer, which also gnaw the bark.

Rodents, hares, rabbits and deer will also damage root crops in a variety of ways. Voles seem to prefer an underground approach, and the Water vole is the star performer. It will eat its way into turnip or swede from below, remaining invisible, and leaving nothing but a hollow shell.

Fruit, on the tree or in store, will be eaten by many animals. This includes wild fruits, such as rose hips, and in the winter it is possible to find old birds' nests in the hedgerows piled high with the debris of such fruits, evidence of the fact that mice or voles have been using the nest as a feeding platform. Foxes will eat quantities of blackberries in late summer, and this shows clearly as dark, purplish patches in the droppings.

Mouse damage to stored apples

Wood mouse and bank vole damage to rosehips and hawthorn berries

Bank vole

Wood mouse

Water vole

Squirrel

Wood mouse

Bank vole

Damage to hazelnuts and cherry stones

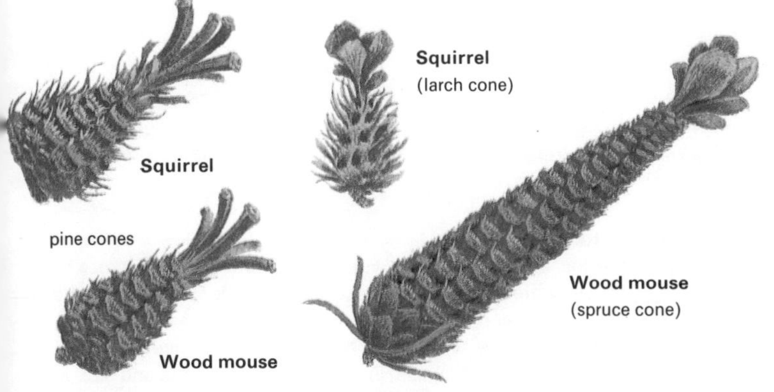

Conifer damage. Squirrel damage to cones is always more ragged than that done by mice

Antlers and horns

The antlers of deer are very different from the horns of sheep, goats and cattle.

Antlers are:
made of solid bone

grown and shed each year

covered with a soft furry skin while they are growing, known as velvet

Horns are:
made of hardened skin resembling finger nails, covering a bony core

grown only once in a lifetime and never shed

hard right from the beginning

An antler shed naturally by a living animal

A hollow, detached horn-sheath can only have come from a dead animal

Antler attached to the remains of a skull showing that the animal must have died without shedding its last antlers

Skulls of horned animals are often found with the horny sheath missing since it comes loose after death and may be separated if dogs have been scavenging on a carcass

The skull of a male deer that has recently shed its antlers

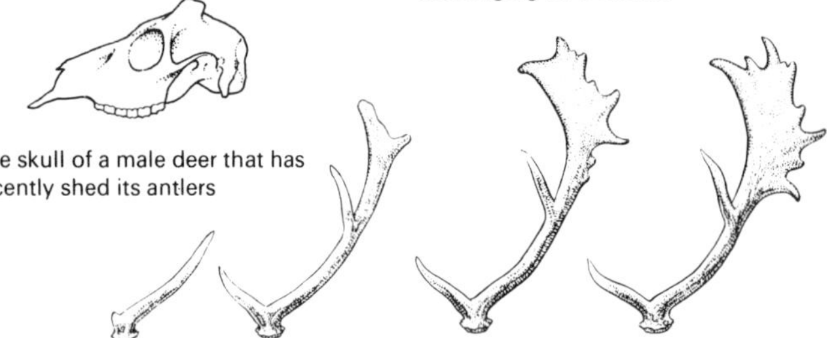

A set of antlers shed by one fallow buck in successive years. The increase in size from year to year is very variable and cannot be used to age an animal accurately, although it provides a rough guide.

Bones and teeth

Whole skeletons, or more often single bones or teeth, can sometimes be found on the seashore, on moorland, in caves or outside the burrows used by carnivores. Skulls and lower jaws are usually fairly easy to identify, especially if they still have teeth; other bones are more difficult. In mammals the teeth in any one skull usually differ greatly in shape and size. In reptiles and amphibians the teeth are generally of one kind.

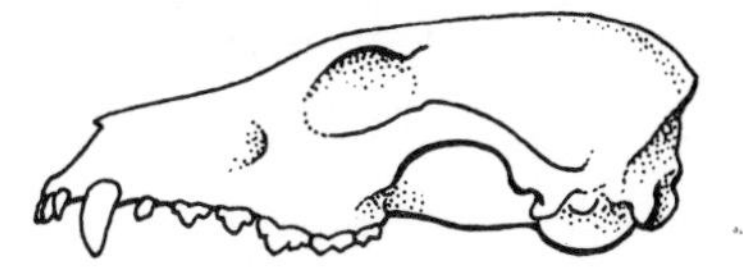

Fox
Note the continuous row of teeth in the jaw and the large fangs. These features are found in the skulls of all the carnivores and seals. Badgers are peculiar in having the lower jaw locked to the upper so that it can swing open but cannot be dislocated.

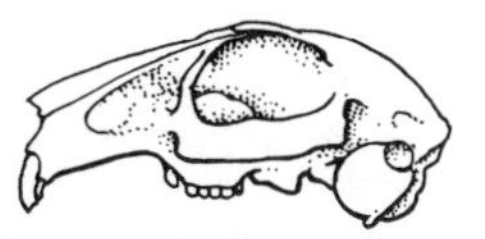

Rabbit
Note the toothless gap in the jaw, the chisel-shaped front teeth and the *second* pair of small teeth just behind the upper front teeth. This kind of skull is also found in the hares.

Mouse
Note the long gaps in the tooth-row as in the rabbit and the *single* pair of chisel-shaped front teeth (usually yellow or orange). All rodents have these features.

◁ **Common shrew**
Note the continuous row of teeth, the long slender jaws and the absence of enlarged fangs.

Lizard
Note that all the teeth are similar.

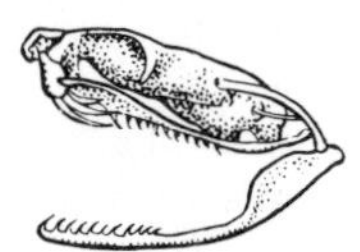

Adder
Note the enlarged poison fangs.

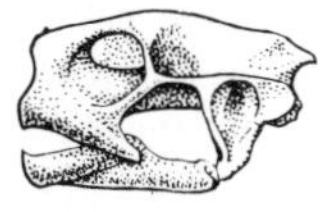

Tortoise
Note the complete absence of teeth and the rather bird-like beak.

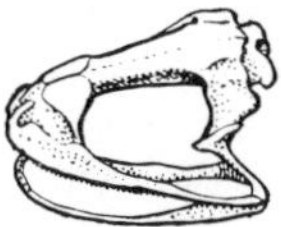

Frog
Note the very numerous tiny teeth and the slender jaws.

Large teeth dug up in a garden or found on farmland or on a beach are most likely to be from domestic animals:

horse

cow

sheep

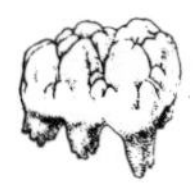

pig

The distribution of animals in Britain

Most of the animals included in this book have very wide ranges, often extending far beyond Europe, in some cases as far as Japan. Within Europe distributions tend to correspond roughly with the major vegetation zones. As a general rule animals that reach northern Europe, like the stoat, pine marten and adder, also occur in Britain which they were able to colonize soon after the Ice Age when it was still connected to the continent. But there are exceptions – some, like the reindeer and lemmings, became extinct in Britain at the end of the Ice Age; others, like the wolf and bear, have more recently been exterminated by man. Ireland was cut off from Britain even earlier and has very few wild animals.

Many species, especially amongst the reptiles and amphibians, are confined to the southern half of Europe and these generally do not occur in Britain. Mammals in this category are the Garden dormouse, the White-toothed shrews (which reach the Channel Isles and the Scillies but not mainland Britain), the Beech marten and the Genet. The higher habitats of the Alps and other central European mountains also have several distinctive species, some, such as the Mountain hare, shared with Arctic Scandinavia and Britain, but others that are peculiar to themselves, such as the Marmot, the Chamois and the Ibex.

Man has not only exterminated animals, he has also introduced them into new areas. The Coypu, Grey squirrel, Muskrat, American mink, Raccoon, Raccoon-dog and Bullfrog have been deliberately brought from other continents; others, like the Porcupine in Italy and the Marsh frog in England, are more local transplants, and in many cases there is doubt as to whether the present range is natural or man-made.

Within Britain some of our more common animals are not only widespread and abundant but very versatile in their choice of habitat. Wood mice, for example, are equally at home in woods, hedgerows and gardens and even on moorland and sand dunes wherever there are a few bushes. The Red fox is another enterprising species, almost as familiar to some owners of suburban gardens in the south as to the Scottish shepherd tending his sheep on open windswept hills. The Common frog, in spite of its greater dependance upon temperature, can be found from lowland ponds to exposed pools high on Scottish moorlands. Many other species however are more restricted in their distribution. Amongst the bats, rodents and reptiles the number of species diminishes as we travel north whereas the larger mammals are least common in the populous southeast of Britain where they have been more heavily persecuted.

Well wooded country in the southeastern half of England, especially where it is interspersed with open country, probably contains the greatest variety of species. Many of these are ubiquitous throughout the country, among them the Badger, Fox, Stoat, Weasel, Hedgehog, Mole, Common and Pygmy shrews, Pipistrelle, Bank and Field voles, Wood mouse, Grey squirrel and Rabbit. In addition there are more restricted southern species to look for, although they tend to be less abundant and more difficult to locate. The Hazel dormouse is one of these, an elusive species found in woodland with a dense shrub layer. Others are the Yellow-necked mouse, which lives alongside the very similar Wood mouse especially in mature woodland; the Harvest mouse, found on the edges of woods where there is tall grass; and the Grass snake, occurring especially in marshy clearings in woodland. There are also several bats, including the Serotine, the Noctule and the Barbastelle, that are confined to the south or southeast. The presence of a river or pond will add several more species to the above list – Water vole, Water shrew and Mink amongst the mammals, Common frog, Common toad and any of the newts amongst the amphibians.

Much of lowland Britain is, of course, farmland. Although the variety of animals is less, many species have nevertheless adapted themselves to the unnatural conditions imposed by fields, especially where remnants of a woodland environment have survived in the form of hedgerows and copses. The mammal best adapted to living in arable land is probably the Brown hare and because it ranges widely over open fields it is less of a problem to the farmer than its relative the Rabbit which concentrates its attentions on the edges of fields within reach of cover suitable for burrowing. Wood mice will occupy growing crops extensively during the growing season but are forced to retreat to the hedgerows and woods after harvest. Hedgerows that contain a variety of shrub species, provided they are not overtidied, are an ideal habitat for all the common, small woodland mammals and for some of the scarcer ones too, especially the Harvest mouse. They also frequently hold House mice and, especially if there is a wet ditch, Common rats whose runs and holes are generally conspicuous on the sides of the ditch. Heavily grazed pasture provides little cover for animals like these but it is a rich habitat for earthworms which are in turn exploited underground by moles and on the surface by hedgehogs and badgers during their nocturnal forays from their homes in adjacent woods and hedges.

Towns and their suburbs are even more remote from the natural conditions to which most animals are adapted but a few species thrive even there. Hedgehogs find playing fields and lawns good hunting grounds and are small enough to need little in the way of shrubbery or waste ground for their daytime nest. Foxes use their proverbial cunning and resourcefulness to survive right into city centres, hunting rats and mice and using railway embankments, factory sites and neglected gardens as refuges and highways. Badgers are less versatile but nevertheless find the outer suburban zone of cities such as London to their liking, using larger gardens and railway embankments for their sets. Besides the expected house mice and common rats, the ubiquitous Wood mouse is the commonest urban rodent, frequently found nesting in the warmth of the compost heap in the smallest of gardens. The smallest of garden ponds is likewise sufficient to attract breeding frogs and toads in spring.

Other amphibians and reptiles do not take kindly to highly disturbed habitats. One habitat that is of especial importance for reptiles and amphibians in southern England is lowland heath. The few remaining fragments, especially in Dorset and Hampshire, are the sole British habitats for the Smooth snake and by far the most important ones for the Sand lizard; they also hold the last Natterjack colonies in the south of the country. Protection of dry heath is essential to the survival of these forms in Britain. It also supports Adders, Viviparous lizards and Common toads although these also extend into a wide range of other habitats.

Wales is marginal territory for most of the restricted species mentioned so far, but it has one mammal that is very much its own, namely the Polecat. Polecats were at one time found throughout Britain. By the 1920s persecution had reduced them to a small remnant in central Wales but they have now expanded to recolonize the whole of Wales and some of the adjacent parts of England. A small remnant population of Pine marten survives in North Wales but to find this elusive species in numbers we must go to the north of Scotland, especially north of the Great Glen where they have expanded from a small remnant, helped by afforestation. The Wild cat is another carnivore that has benefited from afforestation and is now widespread in the Scottish Highlands.

The plantations of exotic conifers that are replacing much of our natural woodland provide suitable habitats, at appropriate stages in their development, for most of our wild animal species. When open ground is first planted with trees it is generally fenced to protect the young trees from rabbits and deer (which can however thrive without doing much damage in mature plantations). The resulting lush growth of grass and other herbaceous plants provides ideal food and cover for field voles in particular whose runways are usually shared by both common and pygmy shrews. As soon as a shrubby growth has developed, bank voles and wood mice move in along with their predators the stoats and weasels. On heathery or sandy ground, the Adder, the only poisonous snake in Britain, may be locally common at this stage, along with the Viviparous lizard. The larger predators – Fox, Badger, Marten and Wild cat – also thrive in these conditions. As the trees mature, only wood mice and a few shrews survive in the deepest shade but all the other species survive in rides and clearings, and red squirrels occupy the canopy, especially when crops of seed-bearing cones begin to appear.

Throughout most of England and parts of lowland Scotland the Red squirrel has been replaced by the introduced American grey squirrel, but the greater isolation of the new Scottish forests gives the Red squirrel a better chance of surviving there. Another species that has survived better in Scotland than in England is the Otter, now very scarce on lowland rivers because of a combination of disturbance and pollution but still present on highland rivers and lochs and especially on the rocky west coast of Scotland where it is as much at home in the sea as in fresh water. Finally the Scottish hills hold our only 'arctic-alpine' mammal, the Mountain hare, especially abundant on the lower heather-clad slopes of the eastern Highlands, but extending right to the highest summits of the Cairngorms.

Deer have a somewhat erratic distribution in Britain because of extinctions and introductions. Throughout Scotland the Red deer is the characteristic species of the open high ground, sometimes in herds of a hundred or more, while Roe are abundant in woodland but much more solitary than the Red deer. In England Roe also occur in the north and the southwest, but introduced Fallow and Sika are the dominant deer in many places. The little Muntjac is much more elusive than the other deer but is now widespread in much of southern England.

Seals can be seen on most British coasts although they are rare in the Thames Estuary and the eastern half of the English Channel. The larger Grey seal is the dominant species on the more exposed coasts from Cornwall to Shetland and south along the east coast as far as Yorkshire. Although breeding beaches are mainly limited to a small number of large colonies they disperse much more widely outside the autumn breeding season and can be seen in the water and hauled out on rocks on most undisturbed parts of the coast. The smaller Common seal can be seen in many of the same areas but on the whole it favours more sheltered waters and is the commoner species in the east-coast estuaries such as the Tay and the Wash, where they adopt a very characteristic 'banana' posture, with head and tail clear of the ground, as they bask on sandbanks.

Seals are equally abundant in Ireland but when it comes to terrestrial animals, Ireland not only lacks most of the more restricted species found in Britain but several abundant and widespread ones as well, including the Mole, Common shrew, Field vole, Water vole, Weasel, Warty and Palmate newts, toads, Slow worm, Grass snake and Adder. Perhaps the most distinctive of Irish mammals is the Irish hare, a race of the Mountain hare that does not turn white in winter and is found on low and high ground alike. Besides more natural habitats, golf courses and airports are especially favoured by Irish hares and sometimes attract exceptional concentrations.

Index of English names

Names in **bold** type indicate species found in Britain

Index of scientific names

Names in **bold** type indicate species found in Britain